MW01253739

HELD TO THE
FIRE

MATTHEW FLAGLER

FIRST EDITION

ISBN 978-1-7389365-0-2 (print)

ISBN 978-1-7389365-1-9 (ebook)

Held to the Fire

Matthew Flagler grew up in Peterborough, Ontario. He graduated from DeVry University with a Bachelor of Science, Business Operations Degree. *Held to the Fire* is his debut book. Matthew's path to becoming a published author began in Grade 10 English when a teacher assigned the book *Firestarter*, by Stephen King as part of the literature studies for the class. After that, Matthew knew he'd be an author. With a voracious appetite, he devoured King's books, then moved on to other horror authors of the day; and began crafting his own stories. Matthew believed his debut would be a work of fiction, but just like the real people in this story, life sometimes doesn't work out the way we hope. And sometimes, part of the process of maturing into who we are, are those trials and tribulations.

Dedication

I dedicate this book to the memory of the women and men employed at the Quaker Oats Company in Peterborough, Ontario on Monday, December 11, 1916. They may have begun their day just as any other, but became heroes before it concluded.

Prologue

Fire has so many similarities to human life: so fragile, but with the capacity to be deadly. Under the right circumstances, both give light, nourishment and warmth yet snuffed in an instant; if either gets out of control, the potential exists to destroy the very things they warm, feed and illuminate. Each carries a kernel of destruction so forceful it can reduce cities to ashes and bring even the most powerful to their knees.

Both breathe, consume, grow, develop, and have the potential to leave a mark on anything they touch. Where the two become divergent is that fire is powerless to be anything but that which it is. Fire may only do; it is. It does not think for itself; fire merely acts. Still, we cannot fault the fire for being what it is or acting as it does. We may only tolerate it or do without. And therein lies our crucible.

This is a story about a fire, yes, but more so of resilience, courage, and bravery. The Quaker Oats fire is something that happened that day. There were many fires before and after. The enduring message of hope, and the human potential to leave one's mark, is the actual story here. There was not one hero, but many. Within the narratives of the legion of individuals who formed part of a larger community and stepped up in the face of adversity, we find the true story here. Not only one of tragedy, but one of hope and humanity. This account is as relevant today as it was a century ago.

My great-grandfather was a man named Dennis O'Brien. He was a Motor-Grinder in the Dry House Area of the Quaker Oats Factory on Hunter Street in Peterborough, Ontario, Canada. Dennis worked in front of a machine that ground raw oats into agricultural feed. I never got to meet him and I didn't know the story of what happened at the Quaker Oats on December 11, 1916, until many years later.

The genesis for writing this book began around 2016 when a local theatre group staged a play called "*The Hero of Hunter Street*". The storyline centred around my great-grandfather, based on historical reports of his selfless heroism on the day of the fire. I loved how they told the story, through the eyes of the people who became embroiled in the tragedy, and how the play paid homage to those who had died. For me, though, it was nothing more than an interesting story to which I had a familial connection.

That same year, a lot of attention locally focused on the upcoming centennial anniversary of the fire, which included publication of a book by local author Gord Young entitled "*A Dark Day In Peterborough*". Other tributes included the release of a video called "*Tragedy on the Otonabee*", then a special issue of the *Heritage Gazette of Trent Valley Archives* highlighting the History of Quaker Oats in Peterborough. A series of commemorations followed, including the erection of a Memorial in early December 2016.

This ruminated in the back of my mind with no actual shape or substance, only the ungerminated seed of an idea. As a proud former member of the Fire Service in Ontario, I have always been interested in stories about fires, especially the fire cause and determination aspects. The W5 of the situation if you will: who, what, when, where, why. A spark that lit the fire (pun intended) was when my Mother and I attended an annual Civic Awards presentation ceremony hosted by the City of Peterborough in the Fall of 2018.

The committee formed to erect a memorial and lead many of the local events were to receive an award from the City. My Mother didn't sit on the committee, but her oldest sister, Doris O'Brien Brick, had. Doris since passed away, and my Mother wished to have family representation at the ceremony. There, I first met Gord Young, and learned of his book, which I purchased out of curiousity and to support a local author.

After reading the book, I undertook research from a general interest standpoint. Not aware of any plan to write a book, the tragedy was just something I wanted to know more about. I found several articles and stories of those who died. As I read more, what at first was an interesting story somehow became more than that.

A lot of information on the mechanics of the fire existed. The layout of the physical plant, timeline of events, and the names of those who played a role in the emergency's mitigation. But I felt a piece of the puzzle was missing. That the story needed to be told another way. I hungered for a more human face to the events of December 11, 1916.

Still, who was I to write this? Yes, Dennis O'Brien was my great-grandfather. Sure, I am experienced and knowledgeable about the Fire Service. OK, I had done a little writing for fun as part of my travels. But I did not see myself as someone who could write this book. For at least a year and a half, I did not write a single word. When it fancied me, I continued to research, nothing more. I am uncertain of when the seed sprouted in the first place, but this scene kept recurring in my mind of Dennis O'Brien, sitting on the edge of the bed, in the cold pre-dawn hours of December 11, 1916.

At first, it was only a still image, a photograph of sorts. That blossomed into bits of dialogue with Laura. I imagined the calm before the storm on that morning, a peaceful scene of everyday life for two seemingly ordinary people. Yet I did nothing more than observe this scene, which began as just described. Over time, that vison fleshed out, becoming something real. Too caught up in my own world, I had time for little more than the observation.

That world changed in mid-March 2020. A global pandemic reared its ugly head, and I thought more about human nature in the face of adversity. Our entire modern lives changed within a matter of weeks. The story of what happened in Peterborough in 1916 once again bobbed to the surface. As I considered what the future may reveal for us of today, a lightbulb turned on for me.

In my free time, I started writing.

While I had the beginnings with the Dennis O'Brien scene, I was not even confident I could write more than an essay, perhaps a short story. Would it even be any good? Maybe no one wanted to read it. I doubted I even had a book in me, since I'd never taken on a project this far outside of my comfort zone. But the words came pouring out from the deep recesses of my mind. What follows is the result.

Things you'll want to know before reading this story. First, it's defined as creative non-fiction, based on true events. Except the fire happened over a hundred years ago. No one is alive today to say within a moral certainty what happened. I gleaned a great deal of information from the hardworking researchers who came before and who published accounts of their own.

But there were gaps. I took a few liberties based on my intuition, common sense and general knowledge of the period. I'm asking you to take a leap of faith too, for the sake of the continuity of the timeline. No one knows, for example, what Dennis and Laura spoke about while enjoying breakfast that morning; nor how Laura felt when she learned of an explosion at the plant. But we can imagine a typical morning routine for a family of the time, or how we'd react when getting bad news.

These were actual people who existed in the fabric of time and space. They had hopes, dreams, and desires for a future that didn't always come to fruition because of circumstances beyond their control. This story is best told within the framework of a humanity. That is how I have told the tale. Yet it needed a little something extra in the telling, and that information was not always available. Everything else is told as accurately and detailed as the information available.

A story of this magnitude does not come to fruition in isolation. I wrote it; it is mine, but I am forever grateful to those who made the job so much easier in terms of the historical facts and context. Foremost, my mother, Julie O'Brien. She expressed I needed to tell this story, and that I was the best person to write it. I am thankful she persisted in encouraging me to write and for believing in me even when I didn't.

A thank you goes out to Gord Young, along with his team of researchers, for writing what the media describes as the most complete story of the tragedy to date. Thanks to Elwood Jones, a renowned local authority on all things history, related to Peterborough and Area. Mr. Jones has written so much on many topics and is a wealth of information. Credit is due for Gina Martin, who researched and wrote of the human side of the fire for the *Heritage Gaze*tte.

My thanks to the Peterborough Public Library, Trent University, Peterborough Museum & Archives, and Trent Valley Archives. These organizations maintain historical records, including the ones I needed to research this project. The digitization of entire newspaper editions and local directories made it easier to do research from home. This was valuable when Covid-19 curtailed the ability to visit these facilities in person. Jon Oldham at the Peterborough Museum & Archives was very helpful.

In closing, undying gratitude goes out to you, the reader, for picking this book up, and for honouring me by allowing me to tell this story to you. I hope you enjoy reading the book as much as I did in writing it. It was a labour of love, rewarding but bittersweet.

MATTHEW FLAGLER
JULY 2020
PETERBOROUGH, ONTARIO

Chapter One

DAWN BREAKS

> *"Life is not meant to be easy, my child; but take courage: it can be delightful."*
>
> - George Bernard Shaw

Dennis O'Brien rubbed his eyes and took a moment to banish the cobwebs from his brain. He glanced out the bedroom window into the darkness, squinting to see what the day had delivered. The room felt like the inside of an icebox. Dennis exhaled to see his own breath and muttered that he'd need to get a fire going. Laura stirred, protected under a fort of quilts.

"*Mm mmm? What time is it?*"

"*About half five.*" Dennis said in a hushed tone.

"*What's it doing outside?*"

"*Looks overcast, might snow today. Cold. I'll head down and see if I can't get that fire lit.*"

As Dennis nursed a fire to life from embers in the bowels of the cookstove, Laura lit a lamp before filling the kettle with water to make tea. Dawn had not yet broken, and the playful glow of the lamplight danced a jig against the nondescript pale walls of the kitchen. They soon settled into the comfortable routine of a workday. Laura made him a boxed lunch as Dennis headed back upstairs to get ready for work.

The pair enjoyed their time together in solitude, before the children woke and got underfoot. Dennis returned to the kitchen as Laura set the table for breakfast. It was a simple meal comprising a soft-boiled egg in a cup each, and dry toast along with black tea. Because of the war, they rationed food: staples like butter, sugar and milk, used sparingly. Dennis and Laura chatted in hushed voices so as not to wake the house. The conversation meandered from the latest picture show playing at the Strand Theatre to whether Henry Burr was a better crooner than John McCormack.

Dennis and Laura O'Brien lived modestly and couldn't afford such luxuries as a moving picture show or to have a phonograph in the house. Though it was still nice to dream about such extravagances. Dennis earned less than $20 a week, and Laura did not work outside the home. They did well to keep a roof over their head and the children clothed and fed. Still, they had accomplished much, considering.

"*We're lucky, though, Laura, when it comes right down to it, aren't we?*" asked Dennis, wiping a bit of egg yolk from his chin before taking another sip of tea.

"*That we are, love. I remember granddad telling me of life back in Ireland as a boy.*" Her eyes moistening with emotion.

Laura spoke of how her grandfather grew up in squalid conditions and abject poverty in the lanes of some non-descript village in County Cork, Ireland. The family had little more than the clothes on their backs and a few meagre sticks of furniture to their name. Muddy streets and the stench of human waste were part of everyday life, along with the underhanded landlords and the lack of prospects.

Dennis nodded in agreement as she spoke. He'd listened to similar stories from his own grandparents. As children, his own parents often went to bed hungry, waking to the cold, hard reality of life in Ireland. Many, born into such conditions, died young. Their desperate situation forced the O'Brien's, like scores of other families, to seek a better life elsewhere. They left the relative familiarity of Ireland to embark on a journey to the wilds of Upper Canada.

The O'Brien's lived in a quaint clapboard house at 552 Harvey St in the progressive town of Peterborough, Ontario. They had five children and Laura was expecting with their sixth, due to arrive in a matter of months. Dennis worked as a motor grinder in the feed room of the Dry House at one of the town's largest employers, the Quaker Oats Company, a few blocks south of where they lived. Laura had an invaluable role in taking care of her family.

After a lull, Dennis and Laura's conversation drifted back to the present. Christmas was two weeks away, though the grim reminder of the ongoing war dampened their excitement of the season. Dennis often worked overtime during the Christmas Season for some leeway to provide gifts.

"*You took an extra shift today, Dennis, and I appreciate that. We'll have a better Christmas now.*" Laura said while clearing the breakfast dishes from the table, "*It means a lot, how hard you work to provide for us. Thank you.*" She gave him a warm smile, leaned in and planted a kiss on Dennis' forehead while tussling his hair.

Like many Peterborough families, the O'Brien's felt the pinch at Christmas, and Dennis had no qualms with finding a few extra dollars. Laura supported her husband's hard-working nature. She knew full well the extra money helped ease their burden. Dennis was thankful to have a job in the first place, at least.

The O'Brien's wouldn't have a transplanted Christmas tree in their home this year. And even if they did, couldn't afford to bury it under a mountain of gifts. Dennis and Laura considered themselves fortunate to have wood to stoke the fire, a modest turkey dinner, and a few decadent desserts. In a good year, there might even be enough rum punch for a toast to the day, and then another.

Christmas was about family and a celebration of life. Which is a good thing, because that's all Dennis and Laura could afford. Their children looked forward to the extra food and treats the most, and at this time of the rolling calendar, Dennis and Laura made an extra-special effort.

Before long, the boisterous chatter of children filled the home. A cheerful chaos as Joe, aged 10, Irene, 9, George, 7, and Kathleen, 6, got ready for school. Michael was three years old and underfoot as the other four scrambled to find their exercise books. They had to be out the door early, with the walk up to St. Peter's School on Reid Street taking at least a half hour. If it took longer, the culprit was always George, known as something of a free spirit. It took little for him to get side-tracked on the way to school and need rescuing from one entanglement or another.

Prior to class, the children attended Mass at the Cathedral. Before leaving for school each day, one of them sometimes asked for money to light a candle at the church. The cranberry red votive holders looked magical with the soft golden glow of a flame dancing inside. There wasn't much extra money to go around, and Dennis often answered *'no'*. This morning, before they even asked, Dennis uncharacteristically produced a coin from his pocket and asked the kids to light a candle and say a prayer for him.

Like many of the working-class families of Peterborough, Dennis and his Father-in-law walked to work. The Quaker Plant ran three shifts a day, six days a week, shut down only on Sunday. The first shift of the day began at 7 am, though cleaning crews started at 4 am on the Monday. William Hogan, a sweeper at the Plant, had left for work hours before.

A 72-year-old retired farmer, William worked at the plant on a part-time basis during the Christmas season, to make additional income. Dennis' wife Laura was one of nine children born to William and Annie Hogan. Since the O'Brien's lived close to Quaker Oats, William found it convenient to stay in town with them during this time. The alternative was to travel back and forth to Smith Township each day by horse and wagon.

The O'Brien's were a loving, wholesome family. Laura and the children depended on Dennis, the sole breadwinner. At around 6:30 am on Monday, December 11, 1916, Dennis kissed Laura and said "*I love you"*. After grabbing his lunchbox and carafe of tea, he walked out the front door of their home for the last time ever.

A brisk wind from the north-east met Dennis as he stepped out onto the sidewalk, biting into his face, and galling him to move forward. He braced himself, lifting the collar on his jacket and putting his head down against the assault.

The ashen sky had a dreary hue, and the mottled clouds suggested snowfall. It was a damp day, even considering the cold. As Dennis turned East onto McDonnel Street, he couldn't help noticing a hint of light struggling to peer through the veil of clouds. The next street over, a dog barked impatiently, no doubt wanting inside. Peterborough awoke from its slumber, and the bustle of activity became pronounced as Dennis passed the courthouse and jail.

Off in the distance, the distinctive clang of a streetcar bell pierced the momentary silence. Dennis imagined it being full of workers headed to the Canadian General Electric factory over on Park Street. The walk to work didn't take over ten minutes, and Dennis often met James Foster walking to the Quaker from his home on Union Street. Today was no exception.

James worked as a stoker in the Boiler House at Quaker Oats. A practical Englishman with the accent to boot, he wasn't afraid of hard work, even at 64 years of age. James, along with his wife Mary Ann, had ten children, including four boys enlisted overseas. Some of their daughters married enlisted men, and one of those sons-in-law became a POW in a German War Camp.

The job of a stoker was a punishing one, requiring James to work in an untenable hot-zone. For the duration of each shift, he fed one of the many coke ovens with coal by the shovelful, standing in front of its gaping mouth and facing a wall of heat. The gases produced in the coal's burning were used to dry the 'green' grain products before milling them into an edible food product, and business boomed.

Dennis slowed the pace as he noticed James approaching, waiting for him to catch up. '*As hard as it is for us, I cannot imagine how difficult*

Christmastime must be for the Foster's, with so many of their sons serving overseas', he thought to himself.

Dennis thought it ironic that, even though World War I caused many hardships, it also meant business boomed for the Quaker Oats Company. They'd received sizeable contracts to produce food for both the Canadian Expeditiary Forces and British Allied Forces. Dennis beamed with pride that he worked at the largest oatmeal milling facility in the Commonwealth and played an important role in supplying the boys overseas with sustenance.

"*Looks like another blustery day in store, doesn't it?*" he asked as Foster caught up.

"*Sure does! This damn wind needs to lighten up, but I s'pose I shouldn't complain, Mesley will have me workin' up a sweat in short order.*"

The "Mesley" he referred to was William Mesley. As Foreman for the coke room of the Boiler House, he was James Foster's boss.

"*How many bushels a day of grain you think we put into those silos, James?*" asked Dennis as they walked south, pointing at the tall grain storage vessels.

"*Mesley says we're runnin' right around 35,000 bushels a day now. Can you believe that? He tells me 40 train-car loads of finished goods leave the factory every day, six days a week. Makes you feel kinda important, don't it?*"

"*Funny, I thought the same thing m'self as you walked up, James. Almost patriotic, if I don't mind sayin' so. We're working hard to keep those boys supplied overseas, and rationing at home to support their effort.*"

Dennis smiled then, pushing his chest up and out while putting both hands on his hips. Foster laughed at the gesture and clapped him on the back.

"*Don't go gettin' your head swelled or nothin'.*" James said with a smile, winking, "*We're 550 strong in the plant and 250 of those are women. We work together. All of us.*"

O'Brien and Foster approached the plant from the north, chatting as they covered the remaining distance, arriving about 6:45 am. A swarm of men and women joined them, hustling toward the Hunter Street entrance. Some of these people had temporary employment during Christmas. Most were full time, a few had not worked in the factory for very long, while others had been employees since the plant opened.

Dennis and James knew most of them, if not by name, at least by faces. Their numbers included both young and old, men and women from many walks of life. They greeted one another with a few words or a smile as they congregated at the front of the building and engaged in a brief conversation as part of their pre-work routine. The latest news of the war was always a hot topic. A mixture of nervousness and fatigue tainted the overheard conversations.

"*How's your son getting along in France? Do you get many letters back?*" asked someone, a look of concern on their face.

"*Did you hear the Germans captured another Allied regiment?*" asked another, sounding worried.

Most knew someone serving in the Great War overseas, and everyone had a heightened consciousness of the important job they fulfilled every day to support the war effort from home. Stories in the news of war tragedies overshadowed daily life, and people were on edge. On Monday, December 11, 1916, the chatter was about something much closer to home.

"*Did you hear about the munitions factory that blew up in Campbellford yesterday?*"

Dennis' ears pricked up, and a heaviness settled into the pit of his stomach as he listened. '*Sabotage?*' he wondered.

The Peterborough Area was an industrial hub supplying the Allied war effort. It produced foodstuffs, electrical components, supplies and many other valuable items for the war effort. For Dennis, it was not far-fetched to think German forces had their eye on Peterborough. His employer could be susceptible to a sneak-attack to hamper production capabilities.

As if on cue, Dennis overheard someone say "*We could be next. First, the Kaiser takes out our bullets, and then he takes out the food supply, yeah?*

I mean, think about it. Quaker is the largest processor of grain in the world. That would be a major blow. I wonder if they'll install air-raid sirens like over there in Britain?"

But even an air-raid siren wasn't going to help them today. In a few hours, a war no one would believe was due to unleash at the factory where they worked.

Chapter Two

A Factory & War

"One of the greatest discoveries a person makes, one of their great surprises, is to find they can do what they were afraid they couldn't do."

- Henry Ford

James Foster wasn't the only employee whose mind weighed heavily on the fate of a family member serving overseas. As Dennis and James arrived at work that morning, Foster spotted Dennis Guerin.

"*Mornin', Dennis,*" James said, greeting him with a wave. "*How's things? Any news about Thomas?*"

Foster was afraid to ask. He knew Thomas Guerin went overseas as a member of the Canadian Armed Forces in the 93rd Battalion.

"*We got a letter Friday.*" he said, with a lift in his voice, and reached into a coat pocket. Dennis teased the well-worn pages from an envelope and began unfolding them. It was obvious to Foster that they'd read the letter many times since Friday.

Guerin cleared his throat and read:

"*The boys say now that our army is top-dog watching Fritzie in the shell hole. We are issued with plenty of charcoal and coke for fuel and plenty to eat. Then we can buy things at the Y.M.C.A. huts to change our menu. For instance, a tin of Quaker Oats and some condensed milk and 10, we have hot porridge in a short time, or vegetable extracts and make some hot soup. We are getting to be some cooks? Well, I should say so. Just wait until we come home and we will show you how to concoct a stew or a pudding and incidentally lower the cost of living. We have hip rubber boots for the trenches and have been issued with new clothing for the cold winter. Au revoir."*

James watched Dennis wipe the tears away as he folded the letter back up and placed it carefully in the envelope. Still clutching the document tightly, when he spoke again, Dennis' voice sounded thick, as if talking through gravel in the back of his throat.

"*Yeah, so I'm glad Thomas is still alive.*" Dennis wiped his nose with a handkerchief. *"You know, he enlisted without telling myself or his mother?"*

James nodded, acknowledging, but remained silent. He stared down at his shoes, not wanting to make eye contact. In fact, James remembered the day the train left the station, on a Tuesday. May 30, 1916, to be exact.

Peterborough was a small community with closely knit social ties, and the Quaker Oats a part of it. Not only did they work together there, but worshipped together and gathered for social events as one enormous family. News travelled fast. From the birth of a child to a new motor car purchase, everyone found out. It wasn't a secret to Dennis' co-workers when his son Thomas enlisted.

Thomas Guerin sat smack dab in the middle of Dennis and Mary Ann Guerin's seven children. Older brother Martin and younger brother Frank worked at Quaker Oats alongside Dennis. Thomas worked briefly there, too, as a labourer before enlisting on January 14, 1916. His sisters, Lenore, Eva, and Hazel and brother John, rounded out the children. Except for Lenore, the Guerin family lived together at 120 Sophia Street, little more than a half-mile from Quaker Oats.

"Dad? Mom? I have to tell you something." Thomas strolled into the Guerin's family living room. He tapped the envelope in his hand nervously on the side table.

Dennis looked up from the newspaper, tilting his head down to peer at Thomas over top of his spectacles. The paper collapsed onto itself for a moment. Mary Ann, who most everyone knew simply as Mary, let knitting rest on her lap.

"*I joined the 93rd Battalion.*" Thomas said, his voice cracking. He didn't dare make eye contact with either. A long moment of silence followed, broken by the sound of crumpling newspaper.

"*You know, Mary, I was just reading in the paper,*" Dennis kept his face concealed behind the newspaper once more, "*the soldiers over there are complaining of defective rifles, boots that rot in the mud, and their trenching shovels are utterly useless. And they're reporting another 300 dead today. What do you think of that, love?*" She could hear the quavering of his voice as he spoke.

Mary picked the knitting back up and looked toward John with a warm smile, then said to Thomas, "*Well. It's a trying time for everyone. Not a simple choice. But, if not my son, then whose, Dennis? Someone has to stand up against this tyranny, stamp it out before it gets here.*"

Early on, support for the war remained high and many local boys signed on, buoyed up by a wave of patriotism that swept through the Nation. However, the war had not ended as fast as expected. As the death toll mounted, enrollment dwindled, and a second wave of recruitment was necessary. For Peterborough, this drive had its beginnings in late 1915, culminating in what became the 93rd Battalion.

The 93rd comprised 500 men, divided into two companies. They recruited members from Peterborough, along with Apsley, Havelock, Lakefield, and Norwood. On May 30, 1916, at 9:30 am, the 93rd Battalion boarded a train at the Grand Trunk Railway station on George Street and headed off to war. The officers marched the entire infantry division right down George Street that morning, past a fanfare of hundreds of well-wishers.

Dennis, Mary, and his siblings saw Thomas off. Mary cried as the train pulled away, amid loud chugs and thick black smoke belching from the stack at the front of the train. The conductor let off two quick bursts of the whistle, cut short by a jarring metallic clank as the wheel pistons engaged and the train lurched forward. Mary hugged Hazel, 16 years old, trying to steel herself on wavering legs as she waved a dainty lace handkerchief.

There were thirty-five pairs of brothers and fifteen father-son sets among the group. A river of tears flowed that morning on the train station platform. The worst was yet to come. Many never saw their loved ones alive again. Few streets or neighbourhoods in Peterborough went unscathed, with 2,300 local citizens enlisting during the war. Over 800 of those made the ultimate sacrifice.

Peterborough was more patriotic than most. The latest recruitment of 500 soldiers brought its contribution to an average of one in nine citizens. That's how many enlisted between August 1914 and May 1916. The National average, just one in twenty-five. Everyone in Peterborough knew someone serving overseas, and it was the predominant topic daily, especially in the Mesley household.

William Mesley and his wife Emma had eight children. Their eldest, Ernie, served in France. Wounded earlier in the year, he arrived home in mid-November. The Mesleys lived in a modest double brick home at 274 Westcott Street, practically identical to the Guerin home. It had a covered verandah spanning the front. William and Emma sat on it during the evening in better weather. They caught up on each other's day, enjoying tea along with homemade scones, still warm from the oven.

All six of the Mesley girls worked at Quaker Oats, too. On days when everyone worked similar shifts, they walked together to work. Emma stayed back to keep the home fires burning while caring for Ernie. On the morning of December 11, 1916, the girls could barely contain their excitement for Christmas. They discussed plans to go shopping after work and even tried to convince William to persuade Emma to let them decorate the house early.

William Mesley had not been with Quaker Oats very long, having been a machinist at the Bonner-Worth Mill. At 52 years of age, he had worked hard. When the opportunity to become a Foreman at the Quaker

Plant presented itself, Mesley jumped at the chance to improve life for his family. His brother Fred put in a friendly word for William with factory management. Fred worked straight days. He oversaw maintenance of the life safety equipment in the plant, including the automatic sprinklers and fire pump.

If truth be told, William had mixed emotions about Christmas this year.

While one of their children died at birth, this was the first year since the war began that the entire Mesley family would be together again at Christmas. He gave thanks their only living son didn't die overseas, a fate that had befallen many other families. On the other hand, William worried they had just enough ahead to make rent for January, let alone splurge by any sense of the word on Christmas.

William felt a twinge of regret that there would be an empty seat at more than a few Peterborough homes this Christmas because of the war. For the moment, he didn't see the irony. In obliviousness, he bantered with and teased the girls as everyone made their way up to the Hunter Street factory for work on the morning of December 11. William loved the girls and beamed with pride as they arrived at work.

When he first came to work at Quaker Oats, the learning curve was steep. Mesley understood little about the cereal business, having worked at a Mill. So, he asked many questions. One of the first, how the factory worked from start to finish. As a supervisor, William had to know the operations in all areas of the factory, not just his own. George Edwards, Assistant Superintendent, took him for a walk-through.

That he was familiar with so many faces as they toured the plant didn't surprise William. He gave a nod or a knowing wave as he passed acquaintances. Even though Peterborough was a city on a growth spurt, boasting a population of 20,000, it remained a small town in many ways. He didn't know all of them by name, of course, but recognized the faces. He'd spoken to them at the weekend Farmer's Market, bid against them at an auction sale, or even sat across from them at the last Church Social.

Edwards showed Mesley where the grain came in by train car for unloading. How machines cleaned the oats thoroughly: chaff, dust, and any foreign materials removed. Only the best part of the oats continued on, to drying, grading, and milling. The oats passed to the next step, separating the "groats" (berries with the hull removed) from any unhulled oats.

After grading the groats a second time for diameter to assure perfect rolling, a process mechanically dried the highest-grade groats. This involved the use of the direct products of combustion of coke. A perforated tube on each kiln directed the flue gases from the burning coke through the oats, drying them to the ideal moisture content for rolling into flakes. James Foster and Thomas Parsons worked to keep the fires hot, so this part of the process went smoothly.

"*Here's where we process the oat hulls through what's called an attrition grinder, William. We make the cereal and animal feeds from the grindings.*"

They'd arrived at the Dry House where Dennis O'Brien worked. As Edwards slid the heavy fire door open, the noise level rose, and he had to lean in to within inches of Mesley's ear and shout a commentary. The steady sound of oat hulls grinding through machinery made talking in a normal tone difficult. Mesley noticed the temperature rise as well.

"*Before entering the grinders, the hulls passed through magnetic separators, aspirators, and other cleaning devices.*"

"*What's that do?*" asked William.

"*Dennis? This is William Mesley. We've hired him as a Foreman for the Boiler House. Why don't you tell him what the magnets and aspirators do?*"

Dennis produced a red checkered handkerchief from his back pocket and mopped his brow with it. Sweat glistened and formed beads on his forehead. He stuck out a hand and smiled at William.

"*I think we've met before. You're up on Westcott St., aren't you? I'm sure I've seen you at Market Hall, over town. C'mon over here and have a look-see.*"

Dennis explained how the safety equipment reduced the risk of foreign objects reaching the grinders. Even a small piece of metal entering the grinders increased the risk of a spark igniting the grain dust. Collection fans connected to the grinders, to keep the accumulation of dust to a minimum. Elevator legs conveyed the finished product into storage tanks in the basement.

Edwards motioned for Mesley to head towards the cereal department next. At the rollers, they watched as William Garvey oversaw sterilizing the groats, and how the machine applied intense pressure, rolling them into flakes. Next, a machine sifted through the flakes to remove any broken flakes and dust before moving along to the packaging area. At the Peterborough plant, almost 200 women, each neatly dressed in a uniform, white apron, and cap, worked to weigh, package and case the finished product for shipment.

Prior to the war, Quaker Oats in Peterborough had a staff of closer to 800, mostly men. Now, despite production moving to three crews operating 24 hours a day, the plant operated short-handed. It had 300 fewer employees on staff, thanks to the war.

Startup for the first shift was underway inside the factory by 7 am. Though the cleaning crews had clocked in earlier, at 4 am, because grain dust is an ever-present danger in a place where oats get processed into cereals and agricultural feed. William Hogan, Dennis' father-in-law, who worked on the cleaning crew, understood that dust is one of the greatest threats to safety in the plant. In the right concentrations, grain dust is highly flammable and bears catastrophic potential.

The factory operated around the clock, six days a week, and everyone took housekeeping seriously. William recalled an incident twelve years before. He didn't work at Quaker Oats then, but in 1904, a grain dust explosion caused a fire in the plant. After, Quaker Oats furnished the machinery in this plant with modern dust-collecting devices such as high-powered exhaust fans and aspirators.

William Denham enjoyed a leisurely breakfast at home on Reid Street as the factory geared up for another busy day. As Plant Superintendent and General Manager for Quaker Oats' Peterborough operations, he worked an office schedule, and arrived at work later than the production staff. Denham moved to Peterborough from Chicago to assume the top position at the factory. He had been with the Quaker for thirty-three years and grasped the business well.

After the explosion in the Peterborough Plant in 1904, the Stuart family, who owned the Quaker Oats Company, hand-picked Denham to take over in Peterborough, the Canadian Head Office. Robert Stuart, both Senior and Junior, continued to oversee world-wide operations daily, from company headquarters in Chicago, Illinois. They'd founded the company after a series of mergers, and before being renamed Quaker Oats, was The American Cereal Company.

The Stuarts purchased land in Peterborough from the Dickson family, along the banks of the Otonabee River, and construction of the factory started in 1901. Known locally as Dickson's Raceway, the land assembly occupied a spit of property from the London Street Dam, south to the Hunter Street Bridge.

The Dixon's had undertaken logging and lumber operations on the property since around 1840, their Mill well known in Peterborough. However, the patriarch had died several years before. So, the business, under direction of the senior Dixon's children, wound down. They were looking to retire and started divesting of real estate holdings, parcel by parcel.

George Cox, a local entrepreneur of some repute, heard the Dixon's wanted to retire. When he learned Quaker Oats had their eye on Canada as part of an aggressive expansion plan because of the favourable tariffs and exporting rules, George wouldn't look a gift horse in the mouth. The factory meant well-paying jobs and positive economic spin-offs for the area. As a business-person, George Cox realized it would be good for his pocket-book, too. So, he acted as an intermediary, and brokered a conversation between the Stuart's and the Dixon's. It worked.

Quaker Oats' locating in Peterborough represented a huge coup for the local economy. Many jobs, paying good wages resulted and the impact, immediate. When the company announced it sought staff, there were long lineups at the door of the hiring office. Before long, a huge shortage of sales clerks existed downtown. Everyone who could, flocked to Quaker Oats for the better wages.

William Denham put safety first. With the rebuilding process in 1904, he recognized they needed to make improvements. So, the Peterborough plant became state-of-the-art in terms of fire safety. Features built into the plant included reinforced concrete (considered then to be "fire proof"), and an overhead sprinkler system. Staff could not smoke within the plant. Each section of the building was fire separated using brick walls, parapeted two feet above the roof.

Automatic sliding, solid wood doors clad with tin protected the door openings between building sections. Pressurized fire hydrants along the east side of the building supplemented the town's hydrants. A dry hydrant installed outside drew water from the river. Many portable chemical extinguishers were available throughout the building, along with water bucket tanks filled with water. The factory even had its own fire pump at the north end of the Boiler House.

William Denham, and the Stuarts, knew Quaker Oats as one pillar Peterborough hoped to build its industrial future upon. They were proud to read that *The Peterborough Times*, in 1906, described the factory as the "*second wonder besides the Liftlocks*". The other pillar being the Canadian General Electric Corporation (CGE). Like Quaker Oats, the CGE was a major employer (400-500 employees), the largest electrical factory in the colonies, and Canadian Headquarters for Thomas Edison's former company.

In the end, Robert Stuart chose Peterborough because of the abundance of raw materials available from the robust agricultural presence in the area. They committed to purchase oats and other feed grains directly from local farmers, as well as from wholesalers and other sources.

The Peterborough factory was massive, with a multi-story brick section at the south-most part of the property facing Hunter Street. This area housed the offices and administration of the company. To the north of that section, Warehouse Number 1 and 2, with the second floor of Number 1 being the Packing Room. These sections were of brick construction, included a basement and four-stories in height.

North of the Warehouses was the Mill Building, also brick construction, and six-equal-to-eight-stories high with a two-level basement. Beyond that, the sections comprising the Dry House and Boiler House. The Dry House featured brick construction with 4-inch reinforced concrete floor slabs on unprotected steel beams and columns. Including the basement, it rose the equivalent to eight-stories in height and overlooked the shorter Boiler House to the north of it.

That Boiler House section was two stories high with a plank-on-steel trussed roof. The aforementioned buildings all adjoined south to north. Next to the Boiler House sat a low, one-story brick building, the Pump House. Thirty-eight feet west of this row of buildings and extending from a point near the south end of Warehouse No. 2 to a point almost even with the middle of the Mill they built a Reinforced Concrete Warehouse.

The concrete warehouse joined to Warehouse Number 2 on the third floor via an enclosed reinforced concrete bridge; and had access to the Mill Building through a concrete tunnel. Behind the concrete warehouse, at a distance of some sixty feet, sat the Cleaning Mill. The Cleaning Mill joined with the Dry House in two locations: via an enclosed concrete tunnel in the top story and via a stone tunnel in the basement. Finally, there were two elevators adjoining the Cleaning Mill to the north.

The factory was a huge, imposing complex. The buildings that comprised the Quaker Plant were as solid as they come. Designed to withstand any calamity, December 11, 1916 would test that theory.

As Denham finished up breakfast, an article in the morning paper caused him to curse under his breath. As he read, William tried to remain calm.

"This reporter has learned about an expected announcement from Prime Minister Robert Borden this morning to issue an 'Order-in-Council'. If enacted, the directive authorizes an increase of Canadian Troops from 250,000 to 500,000 personnel. It represents a 20-fold increase from the original commitment of 25,000 Canadian soldiers made when Britain declared war on Germany in August 1914."

The full ramifications of the announcement dawned on him. While it certainly meant additional profits for the company, it would further tax a threadbare workforce while intensifying trepidation for his existing staff.

"*I wonder if Robert Stuart knows about this?*" he thought, realizing a phone call into Chicago was in his near future. William Denham would call the Head Office in a few hours, but not to discuss the Prime Minister's directive.

A Light,

Well=Balanced Food

In these hurry days we need easily digested food. Avoid an ill-balanced or one-sided diet.

Your food should contain carbon for heat and action; nitrogen for blood, nerves and tissues, and phosphates for bones, hair and teeth. Quaker Oats contains all.

THE EASY FOOD

Quaker Oats

THE WORLD'S BREAKFAST

ACCEPT NO SUBSTITUTE

Chapter Three

A Town's Birth

> *"Be proud of how brave you have been,*
> *taking a risk and following your dreams."*
> - Cassie Mendoza-Jones, "It's All Good:
> How to Trust and Surrender to the Bigger Plan"

"*Peterborough? Never heard of it, Mr. Cox.*" Robert Stuart Senior said. The man to which he spoke was George Cox. While Stuart may not have known, Peterborough was quietly developing a reputation across the Nation. Cox, no stranger to the world of business, knew an opportunity when he saw one, and requested an audience with Stuart about Quaker's expansion into Canada.

George Albertus Cox hailed from nearby Colborne, Ontario. At the tender age of 16, he got his first job as an operator for Montreal Telegraph. Two years later, they made him their Peterborough agent. Three years after that, Cox changed jobs and became an agent for Canada Life Assurance. Cox planned to be the company President one day. Eventually, he owned a majority shareholder interest in Canada Life.

Before long, Cox became a prominent citizen of Peterborough and completed seven one-year terms as Mayor of the town. Cox had an eye for real estate and accumulated large property holdings around Peterborough. By the age of 38, he'd become President of the fledgling Midland Railway. George invested profits from the railway ventures into founding the Central Canada Savings and Loan Company; and later the Canadian Imperial Bank of Commerce invited him to join the Board of Directors.

Cox was part of a group that bought the Toronto Globe newspaper; and of the business association that purchased the Toronto Star newspaper. In 1896, he became a Senator. The public Cox played his cards close to the vest; the private Cox was a simple, kind man who supported many unlucky Peterboroughians and others in need.

Peterborough's other business visionaries, men like Cox's protégé, Joseph Flavelle, along with Richard Hall, James Stevenson, and *Peterborough Examiner's* publisher, James Stratton, devised a remarkable expansion program. By the time the calendar rolled into 1900, they'd laid a solid foundation. Peterborough had the most stable and productive workforce in Ontario and sat poised on the edge of greatness.

"*What other Canadian cities are you considering at the moment, if I may be so bold?*" Cox asked.

"*Kingston, Ontario is on the shortlist. They're your former Capital City, yes? And Kingston is situated right on Lake Ontario. A strategic location makes getting product to market cost effective.*"

"*True, but may I tell you what they don't have, Sir?*"

"*I'm listening.*"

"*Kingston doesn't have Thomas Edison.*"

Cox smiled into the phone. Thomas Edison didn't live in Peterborough. But the founder of electricity located his International company headquarters in Peterborough in 1891, the General Electric Corporation. George Cox was setting the stage for a pitch to Robert Stuart.

"*Electricity isn't the only thing we'd need, Mr. Cox.*"

Robert Stuart knew that the General Electric Corporation had expanded into Canada some years back, but assumed Toronto.

"A factory as large as we're planning will need to access many natural resources. We'll need raw materials to make into finished product. Access to transportation routes to get those goods to market is critical. Yes, we'll need Mr. Edison's electricity, but also a robust and able-bodied workforce, too. Can Peterborough deliver?"

When Cox learned Quaker Oats was looking at Canada, he wasted no time in rallying influential friends and business associates. They met to discuss the reasons Quaker should locate in Peterborough. So, George Cox was ready when Robert Stuart unveiled Quaker's requirements. When Stuart opened the door, Cox outlined how Peterborough was the best choice for Quaker Oats.

Cox told Stuart how Peterborough grew in a few decades, sculpted from the dense hardwood forests that surrounded her. How it was a place rich in timber, water, and other natural resources. That Peterborough had become a valuable economic engine between Toronto and Montreal.

George had first-hand knowledge of how railways cut through the vast undeveloped areas of the territory, linking strategic locations together. He knew Peterborough sat geographically at the centre of a hub connecting a 400-mile network of supply routes, moving immense quantities of natural resources to market. Cox discounted Kingston's advantages by telling Robert Stuart of how the Otonabee River spilled indirectly into Lake Ontario, affording opportunities to ship raw materials and finished goods worldwide.

As important, Cox shared that, because of a series of waterfalls along the Otonabee River, Peterborough reaped the benefits of generating hydro-electric power.

"*In fact, Mr. Stuart, Peterborough is one of the very first places in Canada to generate hydro-electric power, predating even the massive projects at Niagara Falls.*" Cox said, proud of the many 'firsts' in Peterborough, "*and that's not all, Sir*".

Peterborough was the first in Canada to develop a modern gas works system in 1869. It resulted in gas-powered streetlamps throughout the town. In 1884, the budding settlement was at the forefront of electrification in Canada, providing the first electric street-lighting.

"A new project, that's expected to be completed soon, is the Trent Severn Waterway. This major commercial route will link Peterborough to Lake Huron and Lake Ontario. In fact, I'm excited to share that Peterborough is building the largest hydraulic Liftlock in the world, to aid in moving larger commercial vessels between the two Great Lakes. You'd be able to move product from Chicago, directly into Peterborough, faster than if the plant was in Kingston, with all due respect."

George Cox also shared that Peterborough had electric streetcars and incandescent lighting in homes and businesses alike before anywhere else. The production of hydro-electric power and General Electric headquartered in Peterborough were the reason.

"Well, Mr. Cox, you have done your homework, you're saying all the right things here. I'm impressed. Tell me the history of this place called Peterborough you're so passionate about. Is there a stable workforce there? A thriving and vibrant commercial district?"

Stuart was the type of man who wanted to know more than just the dollars and cents of a place. He wanted to appreciate a place for its people as much as its resources. It pleased George Cox that Robert Stuart asked. Stuart was interested. And Cox was ready for those questions, too. He gave the elder Stuart a brief history.

Peterborough had its humble beginnings in 1818, when the "Cumberland" or "Colony" inhabitants arrived and settled at the northern edge of what became the town proper. It was the first large group of European immigrants to settle in the area. A man named Adam Scott arrived in that group.

The following year, Scott built a grist mill at a place the native population referred to as "Nogojiwanong", an Ojibway word meaning "*the place at the end of the rapids*". Water from Jackson's Creek powered Scott's mill. It was at the site of a landfall for a portage used by the Indigenous Peoples along the Otonabee River. A shanty of sorts grew up around the mill and this hamlet became known as Scott's Plains.

Humans occupied the area for millennia. In 1615, Samuel D. Champlain visited during explorations of Upper Canada. He found evidence of early habitation at the sacred First Nations sites of Serpent Mounds and the Petroglyphs. At the latter, stone etchings of daily life, carved by the Algonquin People circa 900-1400 AD, were visible.

Peterborough came into its own, though, in 1825. That's when settlers began arriving from County Cork, Ireland under the direction of the Crown's Land Agent, Peter Robinson. The British government approved an experimental plan in 1822 to populate undeveloped areas of Upper Canada with poor Irish Catholic families. Land Agents such as Peter Robinson facilitated emigration and settlement into various areas.

To qualify for the program, families had to meet certain criteria. They had to be Catholic, poor, and possess a knowledge of farming. Males could not be older than forty-five years of age. Each participant was to be in good health and families not related. Robinson travelled the countryside of Cork, Ireland for three years, selecting suitable families to make the journey.

Immigration Agent Charles Rubidge observed that "*Peterborough grew up as if by magic*" in the wake of the Robinson settlers' efforts. They represented an almost five-fold increase in the local population. Around 2,000 immigrants arrived in 1825, tired and hungry after a month-long voyage across the ocean. William Hogan's parents and Dennis O'Brien's grandparents were among the scores of Irish expats who set foot on Canadian soil for the first time that spring. They disembarked at Port Hope, 30 miles south of Peterborough, on the shores of Lake Ontario. From there, they travelled in caravans headed north.

In 1827, in honour of Mr. Robinson's efforts, Council adopted the proposed town name of Peterborough. For several years, starting in 1845, Sanford Fleming called Peterborough home. Fleming gained notoriety for inventing Standard Time and designing Canada's first postage stamp. He became Sir Sanford Fleming, after being knighted. In 1850, the town incorporated with a population of 2,200 citizens. Many of the original arrivals settled outside the town limits.

Peterborough was an aggressively expanding community. It developed the natural gifts of geography and resources to satisfy the demands of an entire Nation. Between Confederation and the Great War, Peterborough grew up. The economic prosperity that followed exhibited in a rough elegance, characteristic of a town on the verge of greatness. It was common to see grand, ornate houses under construction, against a backdrop of wooden boardwalks, mud-spattered streets and roaming dairy cows grazing on a front lawn.

Industrialization on a large scale changed Peterborough, too. The unassuming little town no one seemed to know, surpassed Ontario's smaller cities. It outpaced others in the value of manufactured products, capital investment, and wages paid to the resultant industrial and service workforce. From a population of around 5,000 in 1871, Peterborough leapt to three times that size in just a few decades.

Peterborough had come a long way since 1819. Many factors came into play. The ability to harness the colossal power of moving water, for one. Abundant stands of timber helped. Access to both vast rail and waterway networks paved the way for growth and a rapidly expanding town market. Settlers from a wide area came into Peterborough to buy goods and services.

Businesses sprang up to service the flourishing demand. Peterborough was the supply hub for an entire region. There were so many new businesses opening along George Street that, before Peterborough was the Electric City, it gained the moniker of the Plate Glass City.

In the end, Cox sold Robert Stuart on Peterborough.

"*My goodness! Peterborough sounds like quite a place, Mr. Cox. You're obviously very proud of it. Tell me, is there a suitable location along the Otonabee River to build a factory?*"

"*Yes, I have just the place in mind, Sir. The land is owned by a long-time Peterborough family, the Dixons. They've operated a mill in that spot for years. But I have it on good authority that the property is for sale. Let me help facilitate the deal, Mr. Stuart.*"

Chapter Four

Red and Blue

> *"True heroism is remarkably sober, very undramatic. It is not the urge to surpass all others at whatever cost, but the urge to serve others at whatever cost."*
>
> - Arthur Ashe

At the Headquarters of the Peterborough Fire Department, staff under the command of Chief William A. Howard began equipment checks just after 8 am. Despite the coolness of the morning, the enormous front doors of the hall sat wide open. A few of the on-duty crew led dappled horses out of the stalls, coaxing them with fresh oats in molasses. An earthy tang of horse and leather mingled with the sweet bouquet of moist hay.

The predominant scent, though, was of fires past, forever embedded into the gear and equipment. A distinct odour, the spicier version of the way one's clothes smell the morning after a campfire, wafted through the station. They thought of the smell as comforting and familiar. One representing victory and success.

Despite the overcast day, morning light coursed into the hall. A corona of hay dust glimmered as it swirled in the air, trapped by beams of light. As the horses nickered, great jets of steam escaped into the bays. Metal shoes clanged noisily on the concrete, disturbing the slumber of the two dogs who called the firehall home, resting on an errant stack of hay in the corner.

The motto of the former volunteer brigade hung on a nearby wall, a reminder that certain truths are eternal: "*Where Duty Calls You'll Ever Find Us.*"

Call-wise, the night before had been quiet. The crew on second watch hoped for the same. By 8:30 am, they'd finished equipment checks and prepared a meal for the team of horses. The animals were how the equipment got to an emergency in haste and worked much better when well-fed. After breakfast, personnel ran a drill to keep themselves sharp, part of the daily routine.

"*Come on, Markwick, double time, double time! Fire waits on no man.*"

Firefighter James Ward barked at recruit candidate Fred Markwick encouragingly, cheering him onwards through the drill. Markwick had been a labourer before joining the Fire Department a month ago and still on probation. Ward cajoled him as a way of making Fred a better firefighter. James saw Fred made a wonderful addition to the team. But he was a probationary firefighter and needed to know they'd set the bar high.

"*Jumpin' Jehoshaphat, Markwick, my mother moves faster when she's burned the stew! Do you want we should have you muck out the stalls again?*"

New recruits ended up with the least appealing jobs in the fire station. The task of cleaning out the horse stalls was one of those. They considered it an initiation of sorts, a test of resolve and willingness to get dirty no matter how difficult the job. Because firefighting was that way. One couldn't '*opt out*' in the middle of an emergency. When the Fire Department made a mistake, people died. Failure wasn't an option, so each took the job seriously.

They called this exercise 'harness drill'. It involved lowering the suspended, ready-to-use tack from the ceiling, hitching the horses up to wagons, and developing steam inside the pump as if responding to an actual emergency. Chief Howard timed the firefighters, who ran the drill multiple times. Everyone took pride in being expedient at their job and teased each other playfully if someone missed a cue or faltered in the order.

After a series of drills, staff inspected the equipment assigned to them. The gear was their lifeline, and Howard required checks after every call and at the start of each shift change. This included a petchcoat, boots, helmet, and gloves. Each man checked their gear for damage or imperfections. Afterwards, either putting it back "in service", ready in case of an alarm, or removing it from service if they found irreparable damage.

A petchcoat was a long trench coat with metal clasp buckles. It served as their personal protective equipment. The coat had a water-resistive coating on the outside, but did little to shield the intense heat of a large conflagration. Their personal gear was basic, and it took real courage to "dance with the red lady" with little more than a raincoat.

"*See these scars, Fred?*"

Ward bent an ear over and leaned in to show Markwick. The ear was red and scarred, damaged in countless incursions.

"*Your ears are the litmus test. When you get a painful burning or tingling sensation around the ears, it's time to get the hell out. I mean it, no exceptions. Do you understand?*"

Ward had taken a real shine to Markwick. As they put the gear away, he continued, explaining that firefighters risk the most to save a life they can save; and a little to save property if they can save it.

"*But we risk nothing… and I mean nothing… to save a life or property we cannot save. It's the hard reality. If we put our life in danger for what we can't save, then there's two casualties instead of one. Do you remember what I told you yesterday?*"

"*Everyone goes home?*"

"*That's right, Fred. Everyone goes home. That's your creed from now on, got it?*"

The firefighters marched into said duties with military precision and took these responsibilities in stride. Lives depended on them being in a constant state of readiness, and the men rarely sat idle at the fire station between calls. Not just the lives of victims at an emergency, either. As a firefighter, each put their life in the hands of another; and asked them to put theirs in your hands, too. Despite having almost non-existent gear, Chief Howard had every confidence in the team. He'd trained them well. Each possessed the determination necessary to face any situation.

The lone fire station in town stood on the southwest corner of Aylmer and Simcoe Streets, staffed by a crew of 16 men. It opened on January 1, 1908 to great fanfare, as the Town prepared to transition to a full-time department. But on June 30 that year, the entire brigade resigned en masse, in protest to shabby treatment by Council of themselves and Fire Chief Rutherford.

The Town didn't invite Thomas Rutherford to apply as Chief Officer and that upset the firefighters. Rutherford, born and raised in Peterborough, joined the Brigade in 1866. He'd served as Chief for 27 years, a tribute to his capabilities, considering the firefighters elected Officers every year. When Council hired an outsider, it was tantamount to insult and to support his crew, Rutherford also resigned.

After a brief recruitment period, William A. Howard earned the top position. The change did little to improve ongoing tensions. Yes, personnel at the Fire Department had altered and the faces of the aldermen different, but there appeared to be an inherent animosity and an air of distrust.

Many in the community had sympathy for the old Brigade. Council didn't rehire most of those former volunteers when the Department transitioned to full time. After Chief Howard's appointment, seven firefighters resigned, perceiving him as an autocrat, and in opposition to his choice of Deputy Chief.

Howard came from Owen Sound to take the position. Known as a steady and practical man, and not inexperienced by any stretch of the imagination. He refused to let safety take a back seat to budgetary concerns. Howard was a vocal, tireless advocate for better equipment and training for the men. But Council considered Howard to be inept about finances.

Chief Howard habitually found himself embroiled in disagreements to justify purchases. Unafraid of controversy, Howard did as he saw best. This was at odds with those clenching the purse strings of the expanding town. The urgency of his pleadings fell on deaf ears and he found no quarter within Council.

Undeterred, Chief Howard pressed forward. He and certain members of Council did not see eye-to-eye on the best way to run a Fire Department. Some saw Howard as careless in respect of authority, though he commanded the respect of the firefighters. This was in great measure because of an uncanny ability to develop a sound strategy for attacking a fire. Howard could put a team in the right place and time to mitigate an emergency.

So, Chief Howard continued to make deputations before Council with requests for new equipment and kept on top of the latest technology available to the fire service. In 1913, Howard attended a Fire Chief's convention in Milwaukee. After returning, he suggested they purchase a powerful motor-driven pump, the latest advancement in firefighting operations.

The motor-driven apparatus developed pressure faster. It allowed for suppression activities to begin sooner. In a line of work where seconds count, this pump represented a tremendous advantage. Considered a frivolity, Council denied the request. Worse, Alderman Johnston, chair of the Fire, Water and Lighting Committee, tried to get Chief Howard fired with a narrow 7-5 margin defeating the motion.

Instead, the Department continued to make do with an aging steam pump mounted to a wagon; one hook and ladder truck pulled by a team of horses; an exercise wagon (used for spare hose) and two hose-reel wagons. The antiquated pump took upwards of ten minutes to develop proper pressure to flow water at a consistent rate. It had to be done at the fire hall before responding to any emergency.

Large industrial buildings like the Quaker Oats plant and the Canadian General Electric Factory were becoming common in Peterborough. Except the Fire Department ladders could only reach three stories in height. Chief Howard requisitioned taller ladders, better hose and equipment. Council denied his requests.

Howard's Deputy Chief was a man named George Jamieson, who'd put in long service to the Peterborough Fire Department. He'd been a member since the volunteer brigade days. Firefighters on duty on Monday, December 11 included Francis Degan, Archie McDonald, George Crouter, Thomas Hatley, and James Ward. These men joined the Fire Department after the collective resignation of brigade members in 1908.

Despite their rudimentary understanding of fire behaviour and the chemistry of combustion, members of the Peterborough Fire Department were ready. What the Department lacked in firefighting apparatus, it made up for in zeal and a commitment to the job.

Around 9 am, Chief Howard told Deputy Chief Jamieson he was leaving the station for W.H. Hamilton's grocery store, on Simcoe St, to pick up supplies. The walk took 15 minutes each way and Howard used the car, owing to the cold. He didn't want to have to carry the supplies all that way back, either. The chance decision to take the car was fortuitous.

Nurses of the Sisters of St. Joseph were well into their day at the 7 am hour. Doctor F.P. McNulty, also the attending physician for the Quaker Oats Company, arrived at St. Joseph's Hospital to perform an appendectomy. The facility had 35 beds, and while busy, was not at capacity.

The flurry of morning activity echoed through the building in the muted tones characteristic of such places. A hushed lilt of female voices set the pitch, while the rhythmic striking of shoe soles on polished concrete floors announced the tempo. These melodies sometimes became interrupted when a loose cart wheel spasmed along its route here, or a bedpan became the victim of gravity there.

The day had begun for the Sisters at 5:30 am. Led by Sister Antoinette, they observed religious exercises in the hospital chapel. After eating, they served breakfast to the patients. As the meals finished up, the nursing Sisters transitioned to bathing patients and general housekeeping duties and at 10 am, the doctors made their rounds.

St. Joseph's Hospital was one of two medical care facilities of its scope in Peterborough. The other, Nicholls Hospital on Langdon Street, just outside the town limits in Smith Township. The Nicholls Family donated the land where the hospital bearing their name was located, on the condition that the town build a Protestant Hospital there.

In order to ensure proper care of his faithful, Bishop Dowling needed to find a suitable location for a Catholic Hospital. He learned five acres of land was for sale between Rogers Street and Concession (now Armour Road) Street. The Church built an ornate solid brick hospital building there in 1890. It stood near the top of Ashburnham Hill, in an area known as St. Leonard's Grove. The west-facing building overlooked the Quaker Oats factory nearby.

Both hospitals in Peterborough found it difficult to recruit staff and attract students. The long hours, poor wages, and demanding working conditions deterred most. It required a robust physique paired with a military-type discipline, but the religious order of the Sisters of St. Joseph attracted many local young women. It afforded them a chance to live their creed of service to God through care of the sick.

The Order had a source of potential staff in the nuns and developed a Nursing School within the local hospital. Their mission: to educate future nurses while providing hands-on training and employment opportunities. The stately building at the top of Ashburnham Hill looked more like an expansive home than an institution. It provided housing for students and nursing Sisters, with a large kitchen to feed patients and the staff/students/residents of the hospital, under one roof.

Even before either local hospital existed, many people associated care of the sick with religion. The Churches, through various religious orders, assumed responsibility to treat the ill. It matched their belief that care of a patients' physical ailments goes hand in hand with care of spiritual needs. The two were indistinguishable.

When the doors of both facilities opened, residents of Peterborough still had to be convinced to come to them when ill. Medical science was rudimentary. Most regarded hospitals as a place of last resort. That changed as training and knowledge improved. Hospitals were less considered portents of death as places that saved lives; even though a few still regarded such institutions with an air of distrust.

For Peterborough, things changed for the better after the two hospitals opened. But Dr. McNulty still remembered a time when there were no telephones or automobiles and few railroads. Money had been scarcer, and medical insurance was non-existent. Births occurred at home. He often made house calls for a variety of ailments. McNulty, like his colleagues, had travelled by horse-drawn wagon and serviced an area as large as a 50-mile radius. It was surprising to him all that had changed in just a quarter century.

In those early days, doctors knew little about bacteria or viruses. The concept of hygiene meant a tub of boiled water and a clean towel. Sterile gauze patches didn't exist, and doctors did not have any protective equipment. For a life-saving surgery, the "operating room" was often the kitchen table, with only a lantern to give light. Antiseptics and anesthetics didn't exist, except perhaps chloroform and carbolic spray.

Natural treatments and therapies were more the norm. Dr. McNulty remembered a colleague sharing how to prepare a medicine by allowing bread to mould and then adding sour cream. The tincture, which they'd called "mould juice", appeared to stop bacterial infections. From local Indigenous Healers, McNulty learned that the tannins in tea soothed burns; and how to make a tannic acid solution for treatment of such ailments. Even four or five years earlier, stories such as one a nursing Sister had told him were commonplace. The story could well have been his own.

"*Once they sent me to Bethany on a maternity case,*" she said, "*the doctor met me with a horse and buggy. The baby had already been born, and the mother was in convulsions. Each time the patient convulsed, her mother, who was there to help me, would faint. We packed the patient in oats, after heating the oats in the oven, and this caused her to perspire. The other children were all sick and one covered with a rash, and her hair was falling out. There were no telephones, so when the doctor made his next visit, he decided it was a case of scarlet fever. He placarded the house for six weeks, which meant we all had to remain there until the doctor officially removed the placard."*

The hard-working nursing Sisters of St. Joseph's Hospital considered the pursuit of a career in nursing more of a calling than a job, unsuitable for the faint of heart. Often the nurses also did manual labour, such as waxing and polishing the floors.

The student nurses attended classes in the evening, after putting in a full 12-hour shift working in the patient wards. After class, they completed housekeeping duties, washing bed sheets and towels by hand, and dried the linens by a fire, before retiring to bed late. This routine started again before dawn the next day. It was a rigorous schedule that tested their resolve.

ATTENTION, MOTHERS!

The famous French physician, Bouchard, says: "*Children fed on meat often suffer from gastro-intestinal derangements, skin diseases and bilious headaches, and rheumatism in its most serious manifestations comes early.*"

FOR INFANTS.—Boil one cup Quaker Oats in two quarts of water for half an hour, strain through a sieve or double cheesecloth, and sweeten to taste.

If you want your boys and girls to feel well—to grow into robust men and women, give them, nay insist upon their eating, QUAKER OATS.

EAT MORE
Quaker
Oats
LESS MEAT

At all Grocers in 2-pound Packages.

QUAKER OATS makes not only the best breakfast porridge in the world, but also delicious wholesome **bread, muffins, cakes, soups and puddings.** Write for our *Cereal Cook Book, edited by Mrs. Rorer.* Free, postpaid.

The American Cereal Co., Monadnock Building, Chicago, Ill.

Chapter Five

A Workday Begins

> *"Every time you wake up, ask yourself, 'What good things am I going to do today?' Remember that when the sun goes down at sunset, it will take a part of your life with it."*
>
> - First Nations Proverb

The familiar aroma of brewed coffee permeated the office as clerical staff filed into work that morning. A rhythmic cadence of typewriter keys hammering out correspondence soon became the soundtrack to another busy day. The telephones were already ringing as Eileen Donoghue took her post on switchboard by 8 am. She routed the mountain of calls that came into the office daily.

While the first production shift of the week got underway at 7 am, administrative staff started their day a little later. Amy Jamieson, a stenographer, had just clocked in. As she hung up her full-length wool overcoat, stuffing her scarf into a sleeve and securing knitted mittens into one pocket, she greeted Eileen. Good friends, the pair had not seen each other since the Friday previous.

Miss Jamieson's twenty-minute walk to work this morning was brisk, most of it with the wind in her face walking north up Water Street. Like many unmarried young women, Amy lived at home with her parents, John and Jane. Her sister Annie lived there, too. Also a stenographer, Annie worked for the Canadian Pacific Railway. Amy worked at the Quaker Plant six days a week and earned three dollars a day.

Amy contributed to household expenses out of her wages. It was part of life, an expectation for anyone who worked outside the home. The only way many families survived was having multiple generations living under one roof, and all paying their share. Still, for a 29-year-old, Amy considered herself fortunate.

Mabel Power soon joined the pair. She worked in the Accounting Department as a payroll clerk. One of many wizards of the accounts, her magic kept the company's finances in balance. The typewriters, telephones and muted swishes of shuffling paper created a symphonic hum through the office. A cough here or a chuckle there punctuated the bustle as everyone settled into their routine.

Mabel's walk into work was much shorter. She lived in a home just across the Otonabee River, at 474 Driscoll Terrace. The front of it faced Quaker Oats on the east bank of the river. Mabel stayed with family, a Rose Power.

Amy Jamieson's manager was William Denham, the General Manager for Quaker Oats' Peterborough operations. Denham's Assistant General Manager was George Edwards. The pair arrived together, amid catching up on the weekend. They discussed the new picture show playing at Mike Pappas' Royal Theatre, called "*The Fight for Paradise Valley*". Shows cost five and ten cents and played in rotation at the theatre downtown, six days a week.

"*Good morning, ladies!*" Denham said. "*Do I smell coffee?!*"

"*Yes, sir, and good morning.*" It was Amy Jamieson. "*Would you like me to get you some?*"

"*I thought you'd never ask! Yes, please. And, Mrs. Power? I need those payroll records on my desk as soon as you get a chance. Are they ready?*"

"*Not yet, sir, but I hope to have that completed before lunch.*" Mabel said.

The women in the factory fulfilled an important and unprecedented role. A decade before, few worked outside the home. Even less in an industrial setting. Peterborough, like other towns, held fast to more traditional gender roles.

But with so many men off to war, an enormous gap existed. Contracts needed to be fulfilled and jobs had to get finished, with or without them. The factory continued to operate, adding a third shift, even. One in nine citizens of Peterborough had enlisted. An equal percentage of the general population were women and children, though. So, not a stretch to imagine one-third of the men in the city had disappeared after the war started.

In the wake of that departure, an enormous opportunity existed. Many girls dropped out of school to take paying jobs at the Quaker plant. The same held true for their male counterparts a few years before. The boys dropped out of school in 1914 and enlisted, willing to make the ultimate sacrifice for their country.

A lot of the women working at Quaker Oats were younger, unmarried, or childless. Ladies like Amy Jamieson, Eileen Donoghue, and Mabel Power. And others, too, such as Eileen Booth, Theresa Dwyer and Eva Tangney. Scores of women reported to the Hunter Street factory six-days a week to fill the human resources void created when the Great War began.

They were as partisan as the men and had a fervent sense of duty. The women saw their work as an important service and became compelled to undertake it. They did so with great pride, unafraid of hard work. The majority worked in the packing room, warehouses, or another capacity in production.

Suffrage was in its infancy. The year before, Canadian women had won a hard-fought battle for voting rights and recognition as a person under law. Much ground remained for them to cover. While the economics motivated the Amy Jamieson's of the company, they also saw this as blazing a trail for those coming after.

Generations of future women owed a debt of thanks to these ladies' vision and boldness in taking a giant leap for their daughters, granddaughters, nieces, and sisters. In a matter of hours, that strength, resolve and compassion would shine. But at just past 9 am on a Monday in December, these ideals were the farthest thoughts from their minds.

Over in the Dry House, operations ran without a hiccup as the shift approached first break around 9:45 am. Dennis O'Brien whistled a tune as the grinding machine in front of him chewed grain into Vim Feed. William Walsh, the crew boss, stopped on his rounds to see how things were progressing. A few weeks ago, a tiny blast in the Dry House Area, known as a "puff", occurred.

A puff is a slight explosion and had become more commonplace. In fact, puffs were so regular that management connected fans to the attrition grinders, to reduce dust accumulation. These fans exhausted into a dustbin on the roof of the adjacent Boiler House. Since puffs could occur in the dustbins, the sides of the bins hung on hinges to prevent such an explosion flashing back into the Dry House.

The seat of most of these puffs was in the attrition grinders, like the one Dennis O'Brien worked at that morning. Although the oat hulls had passed through magnetic separators, aspirators and other cleaning devices, sometimes foreign substances reached the grinders and caused sparks which set the ground hulls on fire. Often the first sign of fire was a distinct and pungent odour given off as the oat hulls smouldered.

On one occasion, such a puff spread fire through an elevator leg into one of the ground hull tanks. A violent explosion happened, one powerful enough to lift the lid of the tank two feet into the air and drop it back into place. Walsh had stopped at O'Brien's station to make sure everything was running like clockwork before calling a break.

William Walsh, 44 years old, had worked at the plant for many years. Highly regarded by all, Walsh held both the admiration and the affection of those who worked under him. A married man with eight children, he lived over on Rogers Street, along with his wife Margaret. One of Walsh's sisters married a Patrick O'Brien so Walsh had a kinship with Dennis.

There was a fellowship among co-workers at the factory, a strong sense of camaraderie. Staff on duty that morning in the Dry House included John Kemp, 66, Joseph Houlihan, 36 and Walter Holden, 33. These men knew each other well, and worked together in harmony, processing oat hulls into livestock feed.

John Kemp was the perfect example of a team player and of service to others. His work comprised keeping the men in the feed room supplied with bags. To the average person, very ordinary work. John considered it as important as any job in the factory. Kemp's co-workers described him as diligent at all times in the discharge of those duties. Although he grieved the loss of his beloved wife on August 18, Kemp showed up every day with a smile on his face, proud to be an important member of the team.

Next door, so to speak, in the Boiler House, James Foster fuelled a massive furnace with a type of coal called coke. He rolled up the sleeves of his shirt to mid-forearm, the cloth itself blackened from coal dust. A river of perspiration spilled from Foster's brow and queued as swollen droplets hanging from the tip of his nose before leaping into oblivion.

The ferocity of the heat was almost unbearable. Foster's sweat-soaked clothing testified to the harsh conditions in which James toiled to keep the factory functioning. Two hours into the shift, with hair matted against his skull, Foster shoveled coke into the appliance, eyes bloodshot from the grime. Caked black rings formed around his nostrils from the coal powder. William Mesley had Foster working up a sweat in no time, as James foretold to Dennis O'Brien on the way into work.

William Miles worked as a stoker in the Boiler House, too. After farming for most of his life in nearby Dummer Township, Miles moved into a house on Rogers Street, once securing the job at Quaker Oats. He and William Walsh were neighbours. At 64 years of age, Miles saw the steady paycheque as preferable to the uncertainty of farming, even if it meant working in such tenuous conditions. The demanding work on the farm, though, such as haying in 90-degree heat with a horse and wagon, prepared Miles for the job as a stoker.

Other people assigned to the Boiler House that morning included George Vosbourgh, Thomas Parsons, Domenic Martino, and Edward "Ned" Howley. Ned had earned the respect of the crew by getting his hands dirty. He often led the charge and just as likely to be hauling wheelbarrow loads of coke to the ovens as it was the most physically demanding job.

As the factory stock checker, John Cunningham kept a finger on the pulse of inventory levels so that supply and demand balanced. John made rounds on the morning of Monday, December 11, becoming uneasy about several things. He noticed an excessive volume of goods in the warehouse, owing to the company's present requirements in fulfilling contracts.

This wasn't the first time John Cunningham observed such conditions, nor the only occasion he mentioned it to management. John still had to report what he saw. Prior to morning break, he made detailed notes on the state of the warehouse.

Most of the stock was extremely combustible and represented an extraordinary "fuel loading". As alarming, the inventory was quite heavy. Starting on the 5th floor of the concrete warehouse building, Cunningham found several wall panels supporting rolls of paper.

On the fourth floor he noted a large quantity of box shooks, rolls of straw-board, and paper rolls. Box shooks were piles of collapsed wooden boxes, used for packing, shipping, or a variety of other uses within the plant. Besides the paper rolls, John saw massive bales and bundles of paper product stored up there on panels, shelving fastened to the walls.

John wrote that box shooks storage crammed the third floor to a state of over-capacity. On the second floor, Cunningham found Quaker Oats in cases, and sacks of both rolled oats and flour. Thousands of large sacks of flour and oats, along with finished product, sat everywhere. At the first floor, John made note of Vim feed, mixed feeds, oatmeal, flour, and more finished product in cases in exorbitant amounts. The basement held Vim feed, along with sundry storage items.

The ideal situation is equal distribution of the stock and raw materials, so the floors could support the load. But on many floors, staff had stacked the product right to the ceilings. Further, on all floors but the 6th, only a meagre ten-foot trucking space down the centre of the floor existed. This put an extraordinary strain on the wall panels holding inventory.

Also, the bulk of the stock on most floors sat in the north end of the building. The load was too unbalanced and concentrated. An imminent collapse hazard existed and needed to be addressed. Cunningham guessed that thousands of bags of hulled oats and flour represented a very heavy live load across multiple floors.

While John did not understand the physics, the architects designed the floors of the concrete warehouse for a maximum live load of 200 lbs per square foot. But each floor above grade experienced a greater load than that, with some carrying as much as 340 lbs per square foot in weight. Paired with the concentration to the north end of the warehouse across multiple floors, John sensed it was a catastrophic failure waiting to happen.

The extreme loads were not the only concern, either. A dangerous condition existed in terms of fire safety. Someone haphazardly piled inventory throughout the warehouse. They stacked it too high, up to the ceilings, with only narrow walkways between piles. If a fire broke out, it would get away on the Fire Department rapidly. Any attempt at fire suppression activities would prove ineffective, and any search and rescue operations, hampered.

John Cunningham finished his survey and left for a scheduled break, concerned with the inventory situation. He also wrote in his notes that he detected an odour of sulphur in the basement, near the ground hull tanks. John wasn't sure, but thought it possible something was smouldering in the tanks.

"If our girls did not eat so much meat, their development would go more steadily ahead."—Dr. Alexander Haig, London.

Children, during the period of rapid growth, particularly require a well-balanced food, such as Quaker Oats, which contains all of the required food elements in their proper proportions. At once healthful, appetizing and economical.

QUAKER OATS is not only the best breakfast and supper dish in the world, but also makes delicious and wholesome **soups, puddings, muffins, gems, etc.** *New Cereal Cook Book* **FREE.**

THE AMERICAN CEREAL CO., Monadnock Building, Chicago

Chapter Six

A Storm Brews

> *"It is during our darkest moments that we must focus to see the light."*
>
> \- Aristotle

Just past 10 am, Mrs. Anne Hopcroft stepped out onto the verandah of her home at 482 Driscoll Terrace. Like the house that Mabel Power called home a few doors down, the Hopcroft residence had a majestic view of the Quaker Oats Plant across the nearby river. As she surveyed the streetscape, Anne's gaze caught unusual flashes of light through the windows of what she believed to be the Boiler House. What followed was a jarring vibration that felt like a large train passing mere feet away.

Before her mind could make sense of the irregularity, the windows blew violently outwards in rapid succession. For one brief second, the melodious sound of glass raining down was distinct above the general din of the town. Then it seemed as if a bomb went off nearby. Anne hit the deck and scrambled to safety.

Ross Leonard Dobbin was one of Peterborough's hard-working civil servants, and Superintendent of the waterworks division. He oversaw the distribution of water through the municipal supply. Dobbin's department, known as the Peterborough Utilities Commission, served the homes and businesses in town and supported fire protection.

The water treatment plant sat along the river, just north of the City limits. But Superintendent Dobbin's office was at 253 Hunter Street, four blocks from the Quaker Oats. From this vantage point, he and a team of engineers monitored water pressure and distribution. A series of gauges and devices allowed them to keep their fingers on the pulse of the underground network of water mains.

Ross took the job in 1914, just three years after graduating from the University of Toronto with a degree in Mechanical Engineering. The Dobbin family moved to Peterborough soon after Ross' birth. His Father was a local historian and Alderman for the City of Peterborough. Dobbin cut his teeth in water management, working on the construction of the Moose Jaw water supply, as resident Engineer for the Walter J. Francis Company.

Peterborough built the original pumping station in 1893. A one and a half story Romanesque Revival style brick building, it sat on the west bank of the Otonabee River. As the demands of the city grew, they scuttled this station in 1909. A salt-box style multi-story double brick structure built downstream replaced it. Both buildings harnessed the power of the fast-moving Otonabee to deliver fresh water to Peterborough.

Around 10 am on the morning of December 11, Superintendent Dobbin and crew were in the office, chatting over a cup of hot tea. The urgent wailing of a low water pressure alarm brought them out of their chairs in a footrace to the instruments nearby. Dobbin watched in disbelief as the gauge for the Hunter Street East Main plummeted to near zero. It recovered towards 40 pounds pressure on the scale and teetered there before continuing a slow climb back towards 70 psi.

Dobbin hammered the face of the gauge with an index finger two or three times, thinking it defective. He opened his mouth to verbalize the thought when the needle plunged back to 40 psi once again. Everyone nearby stood illuminated in the flashing glow of the warning light, watching the gauge see-saw before holding at 40 psi. Clearly something went wrong on Hunter Street.

Dobbin didn't have time to consider what before a tremendous concussion shook the building. He raced towards the front door and out onto Hunter Street, turning to look East. The expression of sheer terror on his face was unmistakable. Ross took the steps back to the office two at a time and lunged towards the nearest telephone. As he dialled a number with trembling hands, time stood still. The annoying rings drew on through the receiver. Dobbin considered hanging up when the sound broke mid-ring, and a voice boomed on the other end.

"*Peterborough Fire Department, Crouter here.*"

After speaking with George Crouter, Dobbin grabbed an overcoat and told head engineer Thomas Park he was going over to the Quaker Oats. He ordered the pressure increased to 100 psi and instructed the remaining crew to double-time it over to the storage garage. Dobbin told them to hitch the dray wagon loaded with supplies and meet at the factory as fast as they could. Dobbin, visibly shaken, rushed out the door.

On the heels of Superintendent Dobbin's call, multiple alarms started wailing in the battery room of the fire hall from the vicinity of the Quaker Oats. First, Box 24, in front of the plant itself, followed in succession by Box 23, at George and Hunter Street, then Box 27 at Water and McDonnel Street. But no one at the Hall needed an alarm to tell them something big happened.

Less than a minute before George Crouter took the call from Ross Dobbin, a loud detonation shook the firehall. They knew something was amiss. But not being their first rodeo, the on-duty crew at the Peterborough Fire Department were already preparing for a full response. Their first order of business, firing up the old steam boiler.

Peterborough Fire Department got early notification of an emergency from "alarm boxes", placed throughout a city at strategic locations, and often on a pole. Such boxes contained either a button, lever, or a telephone-like device, which activated an alarm into the firehall itself.

A mechanism inside the box connected to batteries inside the fire station using low-voltage wires. Each battery corresponded to a specific alarm box location. When a box "activated", an audible alarm sounded in the battery room. Personnel could tell which alarm triggered from there.

From the moment the first alarm sounded, both the team of horses and the resident dogs instinctively leapt towards the front doors of the hall. They too had learned what the sound of the bell meant. By the way the dogs danced back and forth behind the doors, one might think they never got let out. The steeds needed no persuasion to prance forward and into their harness.

A group of firefighters stoked the old steamer pump, persuading a consistent fire using a bellows. Others nudged the wagon forward and hitched it, while the remaining department members snugged up the harnesses. It was a choreographed effort made to look easy by men who had practiced this drill a thousand times. Deputy Chief George Jamieson threw the bay doors open. The team of horses advanced out onto the concrete apron in front of the Hall as the firefighters donned their gear.

The wagons boomed out into the street, one team hauling the steamer and another, the equipment. First, they turned left, and a block later made a right onto Hunter Street. From the rising plume in that direction, Crouter and his men knew where they must go. Amid the boisterous reverberations of bells clanging and hooves pounding, with the steamer belching gobs of thick smoke, the hounds took a strong lead out front.

The job of the firehall dogs was simple. Even though horses and wagons on the streets were commonplace, such a spectacle often attracted the attention of stray packs of feral dogs. It was dangerous for the horses to become agitated by such animals. If the mongrels got underfoot and the team stumbled, it spelled disaster. Likewise, if the horses spooked and bolted, there would be a serious delay in getting to the scene.

Dogs became as valuable a member of the Fire Department as the horses, or any individual firefighter. If any other dogs chased the horses, the department's canines greeted them, along with a vicious snarl or a threatening peck. Sometimes a full out dogfight started. Undeterred, the firefighters continued in response to the call.

The dogs rejoined the team later to watch over their equine friends until safely back in quarters. Many early departments favoured Dalmatians, who were fast on their feet, and taller than most wild mongrels. Though any larger dog, bred to run, did just fine.

Crews working in the Boiler House and Dry House areas of the plant took their break around 9:45 am. They mingled with staff from other areas of the factory, either outside or in a break room. One of the newest members of the staff was a young Italian man named Domenic Martino. He struggled to follow the many conversations in English. Domenic had only recently come to Canada, after hearing stories of untold opportunities here.

Martino had a wife and infant son in Italy. He'd been saving for months to bring them to Canada, and Christmas being two weeks away thrilled him because he'd set enough money aside. In late November, Domenic sent the money back to Italy, along with instructions to come to live in Canada. *We get to be a family again*, he thought, heading for the break.

Martino missed his wife and son. He looked up and noticed two other Italian expats strolling into the break room, Vincenzo Fornaro and Filippo Capone. Domenic's face washed with relief as he hailed the pair with a "*Ciao!*"

Filippo Capone was 20 years old. He came to Peterborough at the request of former neighbours from the old country, the Minicolo's. In a matter of years, the Minicolo family established a thriving grocery enterprise. Capone left Italy alone at 16, en route to Canada. He secured employment at Quaker Oats, since two brothers-in-law worked there.

Filippo, a bachelor, lived with a sister, Carmela, and her family. Their dad, Leonardo, and brother Donato had emigrated to Canada, too, but returned to Italy in late 1915. Ethnic community bonds being strong, Capone may have helped Domenic Martino get employment at the Quaker, too. It relieved Domenic to have someone to talk to in Italian.

Vincenzo Fornaro was the senior of the three, at thirty-one. Capone and Martino may have considered him a big brother, someone older and more seasoned to show them the ropes. Vincenzo and his wife Emilia had two young children, Andrew, nineteen months old, and Mary, four months old.

Vincenzo and Emilia were both new, too, having arrived in Canada in 1913. Fornaro 'aged out' in terms of service in the Great War, but had done his bit with the Italian Merchant Navy. They couldn't have known war loomed on the horizon. But in hindsight, the Fornaro's considered themselves fortunate to be in Canada.

The three "*amici*" enjoyed each others' company during the quick break from work. Other groups, comprising dozens of men and women, caught up with friends and neighbours. Standing in a corner listening, one could catch a loudly spoken phrase here or a hearty laugh there. The overall clamour of the room was that of a school auditorium before the principal arrived, announcing into the mic that everyone needed to take their seats.

A small group of guys who worked together in the Dry House sat in clusters. The timeless adage, '*birds of a feather, flock together*' applied as much in the Quaker Oats as anywhere else. Walter Holden and John Kemp worked side by side in the packing department. Joseph Houlihan nodded to the pair as he strolled in their direction.

"*Hey, Joe, how's Teresa and little Mary? I'll bet she's growin' like a bad weed, eh? She'd have to be, what, nine already?*"

John Kemp asked about Joseph's wife Teresa and daughter Mary.

"*Ten, actually.*" Houlihan said. He smiled and straddled the chair next to them.

"Wow. Not so little anymore. How'd the crops make out this year, Joe?"

"Fair to middlin', I suppose. We'll make do."

Joseph and Teresa Houlihan lived on a farm property in Downeyville. His hike into work was one of the longest of any other employee. But the steady income was worth it, since farming alone rarely provided enough to keep a roof over their head.

Dennis O'Brien poured tea from a carafe into the lid that doubled as a cup, the beverage steaming as if just steeped. Beside him, William Hogan, his father-in-law, munched on a sandwich. Hogan had been on the job since 4 am and it was almost home time for him. The early mornings were hard, but with every ache or pain, he reminded himself it was just two more weeks until Christmas.

As the two chatted, Fred Mesley dropped over to interrupt the conversation. He asked if Dennis knew whether the portable extinguishers used three weeks ago got recharged. O'Brien affirmed he put them back in service full.

Percy Naish arrived for a break. He had been a co-worker of O'Brien's in the Dry House but took a job in the Packing Room last week and wanted to catch up with Dennis. The two discussed the upcoming Municipal election in town. In Peterborough, the election for Mayor happened annually, and as near the end of the year as possible. Naish and O'Brien debated whether incumbent, JJ Duffus, might get re-elected.

Percy was a lucky man. In 1904, he became trapped in the oatmeal section of the plant when an earlier explosion had occurred. The Fire Department had extricated Naish with only a few bumps and bruises. In August 1913, Naish once again cheated death when the Turnbull Building collapsed downtown. Not letting a scare deter him, Percy had even been working as a sales agent for the JC Turnbull Company earlier in 1916, but came back to work at the Quaker later in the year.

Across the room, a group of the English were deliberating the war effort. Among them, Richard Chowen, a native of Devonshire, and a blacksmith by trade. He'd been working at the Quaker Oats for two years. Walter Holden along with James Foster, Thomas Parsons, and

John Kemp rounded out the bunch. All married men, none of them had worked at the factory for more than a few years.

Foster, of course, had the most at stake, with so many members of his immediate family overseas. But the war touched each of the five men, along with many other employees of Quaker Oats. It was an oft-discussed topic of the past two years. No one knew when the fighting might end, but agreed that it could not be soon enough.

The fear and worry over the fate of loved ones was palpable among the general population. One picked up the newspaper with trepidation. They read the updated listing of confirmed dead with one eye open, hoping that might be enough to sponge the name of a brother, uncle or father from the list. No one wanted to, but most felt it necessary. If a family member didn't appear on the list, they expressed a collective sigh of relief; but it was commonplace to find the name of someone they knew.

Unbeknownst to the hundreds of people on break at that moment, deep within the bowels of the factory, a storm brewed. Silent for the moment, an assassin lurked in one of the ground hull tanks on the basement. It waited patiently, needing a breath of air to bring it to life. And as soon as it got that, unleash an unstoppable force. A hint of the impending danger rose towards the upper floors of the building, a taint that revealed the deadliness of the killer.

Chapter Seven

DAMN IT ALL, O'BRIEN, WE GOT FIRE IN HERE

"Fire is the test of gold; adversity, of strong men."
- Martha Graham

There were between thirty and forty men working in the areas of the Dry House and Boiler House of the factory that morning. After break, they restocked the coal supplies, fired up the attrition grinders, and continued with their usual routine for a workday. Ned Howley headed outside with a wheelbarrow to bring another load of coke in from another building.

Jack Conway rushed into the Dry House, and shouted for the Foreman, William Walsh, mere minutes after the grinders restarted. His given name was John, but everyone knew him as Jack:

"*Boss, I'm smellin' smoke over here, good chance we got a puff happening!*"

To Walsh, there was no need to panic, as puffs happened with some frequency. Walsh looked around and spotted his right-hand man, John Kemp.

"John, run down to that elevator boot and have a look see. I'll send Harrison up top to shut off the motors for the grinders until we find out what we're dealin' with here. Grab one of those extinguishers off the wall and take it with you"

"*Roger that, Chief!*" Kemp said, before executing the boss' order.

"*O'Brien! Shut it down. Shut it down now, while we find out what the hell is goin' on here. Spread the word!*"

"*Got it*", O'Brien said.

Walsh located Thomas Harrison and sent him up to the top floor of the Dry House Building to kill the power to the attrition grinders. As Harrison bolted towards the elevator, Walsh stepped over to one of the grinders. He removed the access cover and peered in, squinting to get a better look.

"*Damn it all, O'Brien, we got fire in here.*"

On a lower floor, John Kemp reached the base of the feed elevator nestled among the storage tanks. He set the extinguisher on the floor and opened the access slide in the elevator boot. Kemp watched as sparks ignited the grain dust. There was enough time for Kemp to curse under his breath and register a rush of air from behind, when a fifteen-foot high wall of flames obliterated the darkness. He didn't even have enough time to reach for the extinguisher.

Removal of the cover on the grinder and that of the elevator boot created a perfect environment for the smoldering feed to burst into an inferno. Explosive material in the chase-way ignited. Flames shot straight back up the elevator leg, rocketing towards the Dry House at an alarming pace. A firestorm voraciously consumed everything in its wake, exploding out into the grinder room.

The force of the blast lifted William Walsh into the air like a rag doll. It hurled him out through the windows on the east side of the room, which blew outward seconds before Walsh hit the opening, raining glass onto the ground below. The violent burst propelled his body clear of the Quaker property. He came to rest on the bank of the Otonabee River, broken and bloodied.

A gush of flames ignited John Kemp's clothing, burning everything off from the knees up in a matter of seconds. He staggered backwards into an adjoining department, so obscured in flames that co-workers did not recognize Kemp. As they rushed him to safety, John complained of the cold and asked for something to cover himself. The only items nearby were the very bags that Kemp supplied the men with in the packing room.

Because of open gratings in the floors of the Dry House, the explosion blew the fire into all stories, enveloping the entire building in flames. It didn't take long for thick smoke to shroud the upper floors, and as Thomas Harrison attempted to take the elevator back down, he found no way out. As he scurried back up on foot to the top floor, Thomas discovered the fire had trapped him.

He navigated to the lowest floor safe to enter and attempted to exit through a window. Harrison hung there precariously from an exposed bolt protruding from the brick wall, holding on for dear life. Terrified, he dared not look down. Even though his eyes were closed, Harrison still heard every explosion that followed. He screamed for help as the fire consumed the plant with a gut-wrenching fury.

James Northey had passed Harrison trying to find a way out. Northey found the fire escapes on both ends of the 3rd floor inaccessible. He considered hanging from the side of the building before deciding to jump instead. Northey leapt out the nearest window and tumbled three stories. He hit the ground hard, seriously injuring himself.

Next door in the Boiler House, William Hogan was sweeping around the stokers as he made rounds through the plant. William Mesley was firing up a coke oven just as Hogan passed. In the tight quarters, Mesley closed the door of the oven so Hogan could move past. As he reopened the oven door, the walls came crashing in.

The energy of the explosion in the Dry House blew out both the North and East walls on multiple floors. The North walls bombarded the lower Boiler House from above, obliterating the wood supports and collapsing the roof inwards. William Miles got tossed out a window on the east side of the Boiler House and buried under a mountain of debris.

Thomas Parsons was wheeling a barrow full of coke in the Boiler House. As he tried to avoid a collision with Hogan and Mesley in the tight space, the ceiling collapsed, killing him. James Foster, George Vosbourgh, Mesley, and Hogan died, too, where they lay. The only member of the Boiler House team to avoid instant death was Domenic Martino.

Ned Howley was at the wrong place at the wrong time. When the first explosion rocked the Dry House Area to its foundation, he was outside the building. As he pushed a barrow of coke along the east side of the factory, Howley paused to figure out what happened. The east wall of the Dry House building collapsed, entombing him in the wreckage. Ned didn't even have time to move.

When the North wall fell and crashed through the roof of the Boiler House, it broke the steam connections from the boilers, thus putting the on-site fire pump out of commission. And as the East wall fell, it took out a six-inch sprinkler riser and interred the indicator post-gate valve that controlled the riser, located 26 feet away. In a heartbeat, it rendered the permanent fire safety equipment in the building useless.

The intensity of the release rendered the sprinklers and pumps inoperable, and ripped the fire doors between the Dry House and the Mill Building off their fastenings as if made of cardboard. Unabated, the blast tore doors from their hinges at the far end of the Mill, 130 feet away. Any doors that didn't blow apart were so warped that the automatic sliders could not do their job to seal the rest of the plant off from the inferno.

The ferocity of the firestorm that followed the first explosion raged into a tornado of flames that raced south towards the warehouse areas. Fanned by a moderate northeast wind, it ate its way through the entire length of the Dry House. Dennis O'Brien had survived the eruption. Despite the chaos and mayhem surrounding him, his first thought was to help get the injured out as fast as possible.

James Packenham, a mechanic, had worked at Quaker Oats for years. He was setting up a new machine in the packing department when asked to attend the drying room moments before the explosion. The shock wave following the blast knocked him right off his feet. He had witnessed similar explosions in the past and knew what had occurred.

Instinct kicking in, he ran to give help wherever needed. Packenham found William Teatro just inside the door of the packing room and brought him out. James entered the building again and, on the way to an upper floor, came upon two girls who had fainted. Packenham attempted to enter the building a third time after helping the women, but found the situation untenable and retreated to safety.

The explosion jerked William Garvey off the floor and threw him into a grain elevator where grain buried him up to the neck. A wall of flames that followed drove past Garvey, enveloping his head, burning off the ears and hair, leaving his face a scarred, disfigured mess. It also shrivelled the skin on his head, including the eyelids.

Garvey's brother Frederick, along with a janitor, Edward Skilleter, found him. Frederick was an elevator operator in the plant. They frantically dug William out of the grain while the flames threatened to consume them, too. Neither located a suitable tool for the job, so they used their hands. William, writhing in agony, begged them to hurry. They tried not to look at the mangled features of his head and face, as the stench of burned flesh permeating their nostrils.

William Denham was sitting in his executive office when the first detonation rocked the building. He felt a heavy shock, followed by a distinctive roar. Denham catapulted out of the chair and ran to the river side of the building. As he arrived there, several grain storage tanks tumbled out of the void where the east walls of the Dry House should have been. Denham noted flames enveloped a third of the Mill.

Back at the administrative offices, Denham ordered staff to evacuate the plant, before calling the Fire Department. Mabel Power, along with a few of other girls, gathered record books to take with them. Amy

Jamieson and Eileen Donoghue helped as a chain gang of female staff hauled furniture out the front doors of the building.

Denham noticed this and admonished the girls, reiterating they needed to get to safety.

"*Ladies, please!*" he said, "*the Fire Department is on their way, and they'll get this thing stopped long before it gets to the offices. I need you to get out now, and that's an order!*"

Denham hurried out to offer help elsewhere. The ladies opted to ignore the order and continued gathering the vital company records. Unbeknownst to Denham, fire had ravaged the Mill building and was entering Warehouse Number 1. In mere minutes, the spurious and heartless tendrils of the beast would chew through the remaining real estate to slake its lust. If the girls didn't get the records out now, they'd lose them forever.

Under direction of their supervisor, Miss McCanus, the girls in the packing department were hard at work. What followed the thunderous bang was a fleeting moment of eerie silence. The machinery had stopped working and total darkness ensued as the power failed. If not for the shadows, the faces of the women displayed for an instant a blank, gaping countenance of surprise.

This strange quietness and indecision soon gave way to action. The ladies left the department. They hurried, and with one exception, there was no panic. Many stopped at the cloakrooms for winter coats and boots. A few even punched the time clock on the way out, a system of leave-taking well engrained in them. Regular fire-drills instituted at the factory paid off in that moment.

Miss Eva Booth was the exception. When the explosion happened, she jumped up from her station in the packing department and ran towards the nearest exit. She'd been sitting near a window. Frightened by the gush of flames, Eva bolted down the fire escape before leaping from it and on top of a nearby freight car. She tripped on her skirt and fell to the ground, spraining an ankle. When Eva started screaming, it caught the attention of a soldier, who helped her up and loaned her his great coat for warmth. As she hobbled along, he got her to safety.

Several girls working in the office fainted from the shock of the blast. Mabel Power attempted to return and secure the pay-book of the packing department. She found the way blocked by two such ladies and encountered a third trying to revive them in desperation. Mabel had to leap over the trio to escape, as James Packenham arrived to help the three.

Fire Chief Howard arrived at a scene of absolute bedlam. As he parked a safe distance away, he watched in horror as Walter Thomas Holden ran screaming from the building. Holden looked like a human torch. Someone tackled Holden with an overcoat, throwing him to the ground. Fire had burned his clothing off and singed his hair to the scalp; he writhed in agony. Holden pleaded for relief as the firefighters rolled up to the scene.

Walter Holden worked in the Dry House Area. His job was to sew up bags of feed after being packed. Walter's parents arrived in Peterborough from England three weeks before to spend the Holidays with Walter, his wife, and their two children. It was the first time his parents met their newest grandchild, just 14 months old. Walter Thomas Holden, 33, did not survive to see Christmas.

Company employees worked to rescue the girls in the Package Department on the second floor of Warehouse 1. Under Denham's direction, a team attempted to close any accessible fire doors. Still others had begun a foot search for victims stranded in the Dry House.

"*Jamieson. I want you to put together a team and establish a water supply into this scene. Then put these men to work assisting with rescue operations. I see a man hanging from a bolt high up on the wall over there. See if you can get him down.*"

Howard's instincts kicked into high gear. In an instant, he'd sized up the scene and knew what needed doing. The Fire Department's priority was preventing further loss of life. They could replace a building, not lives.

The suppression crew of the Peterborough Fire Department struggled to establish a water supply. The hook and ladder team joined Quaker staff to affect a rescue operation of the injured. First, they attempted to gain entry into the factory from a metal staircase on the south side of the building. Their assignment, to reach the north end of the building and rescue anyone along the way. They didn't get far. Dennis O'Brien met them coming in and gladly lightened his load by handing off Richard Healey. Healey was unconscious and bloodied, but still breathing, even if it was raspy and uneven.

The side of Healey's head was misshapen and had an obvious indentation along the part of his hair. O'Brien had Joseph Houlihan, too, one of Dennis' co-workers from the Dry House. Houlihan was severely burned and couldn't walk under his own power, but managed with O'Brien's help. Shock had set in, and Joseph couldn't speak. His clothing was in tatters and face blackened.

Dennis assisted the firefighters in getting the two victims to safety. Before anyone noticed, O'Brien disappeared back into the raging inferno. He needed to find William Hogan. Dennis shoved aside a fallen shelf here and stacks of box shooks there. He explored further into the factory, shouting Hogan's name above the snarl of the fire.

The intensity of the heat was something Dennis had never experienced. The fire itself was equal parts terrifying and eerily beautiful. Mesmerized, Dennis watched in awe as the flames licked along the warehouse ceiling. They appeared to dance in thin air, rolling over on an invisible current. A torrent of orange waves gushed in slow motion above him. To Dennis, the room was silent, and he stood slack-jawed, frozen in time.

The deafening report of what sounded like cannon-fire snapped him out of the trance. Back in reality, Dennis raced forward against the ticking clock. He was oblivious to the toll that the heat and exertion were taking on him. O'Brien stumbled upon Percy Naish, the lucky man who survived two close shaves with death. Naish was light-headed and overcome by smoke, but conscious and alive.

Dennis helped him up and hauled Naish towards an exit. After handing Naish off to the nearest help, Dennis entered the building for the final time. As he searched for Hogan, O'Brien was last seen alive from a few hundred feet away by two other Quaker employees. Dennis was on the sixth floor of the Dry House building. One employee shouted to him:

"*Dennis! It's time to go... c'mon, we're gonna die in here if we don't leave now!*"

No one can be sure whether Dennis even understood, on account of the deafening blare of the fire. It's possible that he thought he had one more go in him. For sure, Dennis thought of Laura and the kids in those last moments. The closing words were barely off the lips of his co-worker when the floor disappeared below O'Brien. As a disquieting look of surprise flushed onto his face, Dennis hung suspended in mid-air for one instant before disappearing into the gaping jaws of the void. They never found his remains.

One of the two men would have nightmares about the scene for years afterward. He'd witnessed the last moments of a man's life. He wondered if Dennis' death was agonizing and slow, or painless and quick. For the moment, the two looked at each other in disbelief and ran for safety.

PURE
QUAKER OATS

Chapter Eight

Initial Response

> *"What matters most is how well you walk through the fire."*
>
> - Charles Bukowski

Outside, things were not getting any easier for the Peterborough Fire Department. As the first team entered the building and encountered Dennis O'Brien, another crew conducted a ground search of the exterior. It didn't take long for them to find Thomas Harrison hanging from a bolt on an upper floor of the Dry House. Harrison had been there for almost twelve minutes and about to lose his grip. But the Fire Department ladders only extended up three stories.

As Department tradition followed, they produced a deck of cards. Whoever draws the highest card had to attempt the rescue. After shuffling the cards hastily and fanning them out face down, Deputy Chief Jamieson thrust them forward. Each took a card and showed it. George Crouter drew high card and became the "lucky" man who had to scurry up a ladder as high as possible and grab Harrison to pull him to safety.

Richard Chowen was hoarse from screaming for help. A roaring fire made every conversation a shouting match and, buried under a mountain of debris, no one heard him. As he begged for help from mere feet away, his voice gave out. The weight of the brick tomb that surrounded him, an immovable force. Chowen could not get free. Frightened, the air pressed from his lungs while rescuers stood mere steps away.

The suppression team was in trouble. In an instant, the explosion deprived them of the use of the Quaker's own fire pump. It severed the 6-inch sprinkler riser in the factory and fused most of the sprinklers in the lower floors of the Mill. The street mains leading to the plant could not maintain any pressure. While flames shot hundreds of feet into the air, the Fire Department struggled to establish a water supply.

Superintendent Ross Dobbin arrived on scene not long after, and the Fire Department looked at him to help work around the present challenge. He tabled the possibility of drawing water from the Otonabee River, but Chief Howard dismissed it. The Fire Department's steam pump could not draught water from that height. And the pump was too cumbersome to lower to the river to maintain a prime.

Dobbin dispatched Thomas Park to shut the valve on the Hunter Street end of the yard main, to improve pressure. It didn't. With his mind racing like a rabbit being hunted, an idea came to him.

"*What if we closed the other valve in the yard main? The one just west of Elevator A. That would increase pressure.*"

"*But what about the sprinklers?*" asked Denham, worried.

"*What about them? They don't seem to be doing much good. Do we know if they're even operable?*" asked Dobbin, realizing they were running out of time.

"*We're not closing the valve. If the sprinklers are working, then we have to leave it open. What do you think, Chief Howard?*" William Denham had regained composure.

"*We need to go on the assumption they're still working, at least for now.*"

The Fire Department owned a smaller steam pump, stored at the old Ashburnham Fire Hall. A group went to fetch this pump, hoping they could either lower it to the river to draw water or use it with the dry hydrants installed on the Quaker property. These "dry" hydrants connected to the river through metal standpipes along the east side of the property.

Meanwhile, five firefighters attempted to gain a foothold on the fire.

"*Ward! Mount up, your team is going in.*"

"*Sir, yes, sir.*"

James Ward assembled a crew of four, including Markwick, who looked nervous. This was his first big fire. Ward tapped on his own ear and winked at the probationary firefighter. It took a minute for Fred to register what he was getting at. '*A painful burning or tingling in the ears means it's time to get out.*' He heard Captain Ward's words in his head and fought the urge to vomit. But he soldiered up and followed his leader towards the building.

"*Remember, men. This girl is a fickle mistress. She'll kill you if given the chance. Never turn your back on her, not even for a second.*"

With those last words of advice, Ward turned, hammered the door handle with the head of an axe, and swung the door open. The sound of rushing air when the vacuum of the building envelope broke was like nothing Fred had ever heard. From deep inside the building, it sounded like the fire laughed. An evil cackle that echoed through the dark space beyond the doorway, as if it had been waiting for them.

"*Hurry up, ladies, it's go time!*"

The crew gained entry to the structure on a lower floor. Ward and his team advanced up a stairway, carrying a semi-flaccid line of hose with which to knock down the fire. The heat was intense and, with insufficient pressure in the line, they were back on their heels from the outset. The fire growled at them, furious, showing its teeth and taunting them to step closer.

A few minutes later, Chief Howard entered the building to determine how the firefighters were making out. He found them pinned down by the fire and obscured in a haze of smoke. Howard at once blew a whistle, signalling retreat. The heat and smoke became too much for Markwick, and he collapsed. Howard and Ward dragged him to safety before placing him on a wagon heading to hospital. The rest barely escaped with their lives.

After a brief discussion with Deputy Chief Jamieson, Howard decided an interior firefight was no longer a choice. The Fire Department regrouped for a defensive attack and protected exposures in the surrounding neighbourhoods. Water pressure hadn't improved, and it looked like even saving the building was out of the question.

Superintendent Dobbin took this more personally than most. Despite the earlier order to leave the one valve open, he realized the sprinklers had to be out of commission. On his own authority, Dobbin attempted to locate the valve. He knew where it should be. But there had been construction work completed on the grounds since the last site visit, and Dobbin didn't find the indicator valve post.

He seconded two waterworks employees, deciding instead to close the valves at both ends of the Hunter Street main. This would, of course, cut off the supply of water into the building, but improve pressure in the hydrants and mains throughout the City. Dobbin knew it was the only way to give the firefighters a chance to gain the upper hand on suppression activities of adjacent properties.

Just past 10 am, the students at both King George Elementary and Immaculate Conception schools were busy in class. As teachers gave instruction within the tranquil environment of their respective institutional settings, the looming devastation was unfathomable. When the first detonation occurred, the force of the blast spider-webbed most of the west-facing windows at King George Elementary. The building shuddered. Any classes in rooms facing the Quaker Oats had a front-row seat to the destruction. Few wanted them. Children cried openly, and teachers felt faint, shocked at the sight of the wall of flames on the other side of the river.

High up the hill and oriented east to west, the school felt the brunt of the blast. The building rumbled from the foundation upwards. Fearful of structural damage, officials evacuated the school. Students assembled in the parking lot on the east side of the school. Most only had enough time to grab their outerwear and don it after getting outside to safety.

As with most everyone else in Peterborough, many students and teachers at both King George and Immaculate Conception schools knew someone working at the plant. Students wondered if their mother or father remained inside the inferno. The scene unfolding before their eyes was surreal, and most couldn't fathom the personal tragedy they witnessed.

At Immaculate Conception School on Mark Street, the second-grade class was in an upper story, west-facing classroom when the explosion happened. In unison, students and teacher alike turned to see a cloud of thick black smoke rise out of the building in the area of the Boiler House. One pupil, Pete McGillen, thought a locomotive blew up.

Everyone cringed in horror as flames spewed out of the east side of the factory and sprinted along inside the plant, swirling like a tornado and gaining in intensity. The teacher dashed to the windows and pulled the shades as the school principal came in to ask for a moment of her time. The two spoke in hushed tones outside the classroom door. A few students went to the windows and peaked through the blinds.

A short time afterwards, the teacher returned to the classroom. She found a group of boys spying at the spectacle. With a yardstick in hand, she encouraged them to take their seats. Once everyone had settled, she made an announcement:

"*Class, I just spoke to the Principal. He's cancelled morning recess. However,*"

A collective groan from the student body cut her announcement off mid-sentence. She cleared her throat and continued, trying to sound calm.

"*However, we're calling an early lunch hour. But you're expected to return to school as usual this afternoon. And you are to go straight home. Is that understood? Do not, under any circumstances, go across the Hunter St. bridge.*"

At St. Joseph's Hospital, the boom sent tremors through the building. Dr. McNulty was finishing an appendectomy when it happened. A nurse in the Operating Room noticed flames gushing out of the factory building and alerted Dr. McNulty. He left the remaining suturing to his assistant and made a beeline for the main floor. Whatever had happened, it sounded serious, and Dr. McNulty expected casualties to arrive before long.

Mother Antoinette dispatched a group of nursing students to check on students and faculty at the two nearby schools. Split up into teams, they gathered their overcoats and headed in respective directions. Even from that distance, the roar of the fire was deafening. For many of the young girls, this was their first experience with a crisis and they were frightened.

Fire surged angrily from the factory across the river. From the nurses' higher vantage point, the firefighters appeared insignificant against the unleashed inferno towering above them. The sound of explosions, breaking glass, men barking orders, and the screams of the injured raced across the Hunter Street bridge to meet the nurses. As they marched to the two schools, the bitter wind in their face, adrenalin coursed through their veins. This was real and happening now.

At the schools, the nursing students found no one injured. School officials had things well in order, as the students readied for an early departure to their own homes. After returning to the hospital, Mother Antoinette dispatched the student nurses to attend the Quaker Oats property. She gave orders to aid in any capacity, after reporting to the doctor in charge. Florence Nightingale would have been proud, watching the young ladies trudge across the Hunter Street Bridge in crisp white uniforms and smart navy overcoats. They arrived at what resembled a war zone.

For many miles in all directions and into the countryside, people reported hearing explosions. Within the rudimentary social network of the day, word-of-mouth, the salacious news travelled as fast as the fire itself. Many dropped everything and headed to the factory. Some wanted to witness the conflagration, while others intended to offer help. The well-being of a loved one working at the factory that morning concerned an equal number of residents.

One of those who arrived at Quaker Oats was Dr. Jessica Birnie. Her home and medical office were at 501 Water Street, a third of a mile away from Quaker Oats. When the house in which she boarded trembled sometime past 10 am that morning, Dr. Birnie stepped outside to investigate. It wasn't hard to miss the plume of thick black smoke rising skyward near the Quaker Oats plant.

Dr. Birnie represented a shining example of the fundamental change occurring in the area. A single woman, she was the only female in Peterborough to own a motor car. Dr. Birnie was the first female doctor in Peterborough and one of only a handful in all of Canada. She displayed a penchant for wearing dressy pant suits typical of business executives and not afraid to take charge or bark orders when necessary.

Diminutive in stature and possessing strong features, some described Dr. Birnie as being plain. Though all agreed, she possessed broad common sense, along with a practical approach, combined with just a hint of dry humour. Dr. Birnie was a loved and respected member of the community. Much of the time, apart from professional duties, she devoted to causes for the City's youth and those involved with social welfare supports.

Jessie worked, for example, to set up Maternity Insurance in Peterborough, and the City's first "Well-Baby Clinic". Her academic pursuits included founding a Shakespeare Club in Peterborough in 1912. After graduating medical school at the University of Toronto in 1898, she spent two years at the Polyclinic Hospital in New York studying the diseases of women and children. Dr. Birnie arrived in Peterborough in 1900 to begin general practice.

Not only was Dr. Birnie unique as the Peterborough's sole female doctor, but as a guest lecturer at St. Joseph's Hospital School of Nursing. Topics of instruction included Anatomy and Viscera of the Thorax, Abdomen and Pelvis. For the advances made in gender equality, there were still challenges she faced. One of the most glaring, that the City refused her hospital privileges at Nicholls Hospital for eight years until 1908.

Dr. Birnie was a contrast unto herself. Besides the traditional medical training and expertise to which she ascribed, Dr. Birnie was a fervent believer in alternative medicines and treatments. She often consulted with Indigenous healers when either mainstream medical treatments were not effective, or as a supplemental treatment.

No one knew why Jessie chose Peterborough to establish a medical practice, as she hailed from Collingwood. But the men and women working at the Quaker Oats on December 11, 1916 were glad she did. As she pulled up to the scene that morning, it relieved many to see her. Amidst the chaos of the situation, Dr. Birnie's presence acted as a calming agent.

She sized up the emergency as if alone in command, possessing an innate ability to shut out the distractions and turmoil to focus on what needed doing. Despite a brisk demeanour, locals described her as gentle with patients. Some 200 girls, lost lambs on the grounds of the Quaker factory that morning, were thankful. Dr. Birnie acted as a steadying shepherd, guiding them towards a sense of normalcy.

Dr. Birnie set up triage on the south side of Hunter Street in front of her car. The nursing students from St. Joseph's Hospital congregated there, awaiting orders. They were familiar with Dr. Birnie and gravitated towards her. The female staff of Quaker Oats saw her as a reassuring face, too, and began composing themselves as best they could, considering the shock they'd just survived. Before long, Dr. Birnie had devised a strategy.

"*Girls. Yes...the both of you. Get over here and help me, please and thank you.*"

Dr. Jessica Birnie barked orders like a Drill Sergeant and the student nurses stepped into line without hesitation. It eased their anxiety to not have to think, but to just act.

Chapter Nine

FORTITUDE AND SORROW

> *"That is fundamentally the only courage which is demanded of us: to be brave in the face of the strangest, most singular and most inexplicable things that can befall us"*
> - Rainer Maria Rilke, Letters to a Young Poet

Up at 552 Harvey Street, Laura O'Brien felt the house vibrate and noticed glass chattering in the windows as the aftershock of the explosion reached the neighbourhood. She gathered toddler Michael up into her arms and stepped out of the house, mesmerized by the rising plume of smoke to the South. At that same moment, she caught movement out of the corner of an eye and, looking right, noticed John Todd sprinting past the front of the house. John lived a couple doors away on Harvey Street and worked at the factory.

"*Laura, have you heard?*" he said, out of breath. "*The Quaker Plant just blew up!*"

Fear settled into the pit of Laura's stomach as a serpent cinching around its prey. She stared at the smoke before returning into the home she and Dennis shared with their family. In that moment, an overpowering desire told her to run down to Hunter Street. But the children came home from school at lunch, and she wasn't about to just up and leave.

The O'Brien's did not have a phone in the house. In the Peterborough of 1916, phones were the purview of businesses and wealthy households. Even if they had a phone to call Quaker Oats, there would be no one at the other end to answer. Laura tried to keep busy so as not to think about it, but with each passing minute, her anxiousness grew. At first, she did a passable job of convincing herself that Dennis and her father were fine.

This far into a pregnancy, Laura's emotions were raw on a normal day. But as she imagined the worst, the dread grew to become all-consuming. Slumped over in a chair, face buried in her hands, silent tears fell. The potential impact of it hammered down on Laura like steel on an anvil. Her foreboding only worsened as she recalled Dennis' request to the children that morning, to "*light a candle for me at Mass today*".

So Laura did the only thing she could. She prayed. Because for now, there was no one to call and nowhere to go. In a few minutes, Laura regained her composure and held out a sense of hope that Dennis and her father returned safely home.

The scene on the grounds of the Quaker Oats factory was chaotic. While those such as Chief Howard and Dr. Birnie attempted to wrestle control from the confusion, it was a losing battle. Very few present that morning had ever experienced such a catastrophe before, and panic grew by the minute.

The fire had a voracious appetite and swelled with such ferocity that most people froze in fear. Spellbound, the exceptional power the flames wielded held them in a vice-like grip. With water barely dribbling out of the hose lines and a swirling wind to fan the flames, the situation became dire. Fire brands were being carried farther afield, and there was the real danger of this incident levelling the entire city.

The smaller steam pump from Ashburnham Hall arrived on scene. Firefighters worked together to move it through the minefield of debris up the east side of the Quaker property. If they accessed the dry hydrant near the river, the team might get water moving through the hoses. With great difficulty, the firefighters manoeuvred the pump by hand, but falling walls and debris trapped them. They abandoned the pump where it stood.

The speed with which the blaze gained strength was incredible. Emergency personnel were unable to execute any plan of attack ahead of it, forcing them to act defensively. For all their training, the firefighters did not have the proper tools for the job, nor the luxury of time to formulate an alternative course of action. The student nurses didn't have the proficiency to triage on their own. And it caught City staff comprising the water and power divisions off-guard.

From the first explosion, everything happened in less than twenty minutes. In another ten minutes, the chemical tanks started exploding, echoing muted detonations within the building. Then the floors started collapsing. Overcome with emotion, General Manager William Denham looked for someone upon whom to vent frustration. Scrutinizing the scene, he noticed Chief Howard standing on Hunter Street, arms crossed and brows furrowed.

"*Are you not going to do anything about those explosions in the plant? Are you seriously going to just stand there and do nothing?*" Denham barked.

Chief Howard regarded Denham with pity and was mustering up an answer when City Engineer Roy Parsons interjected instead.

"*That's the floors falling, Mr. Denham, Sir*" he stated, leaving a pause before putting emphasis on the "*Sir*". Parsons turned away, making eye contact with Chief Howard, and winked in a knowing way that said he'd been on the business end of a tongue-wagging himself.

Denham stared slack-jawed at the pair for a moment. If looks could kill, they'd both be laying on the ground. He said nothing, but turned on a heel and stormed off in the opposite direction. The floors were collapsing. As John Cunningham observed, they'd overloaded the warehouse. The intense heat further deteriorated the load-bearing capability of the floors. Even with an adequate water supply, the outcome was unavoidable.

Fire Chief Howard realized minutes before that the entire city was at risk if he didn't get more resources. He sent Deputy Chief George Jamieson to secure a phone line anywhere and call in Lindsay Fire Department to help. With only 16 men, including the injured Fred Markwick, Howard was out-manned. The firefighters dug in, knowing they had to go the distance on this one. Their only hope, that reinforcements arrived before it was too late.

As Chief Howard surveyed the scene, he thought how much this desperate setting was similar to one a group of Canadian soldiers faced at the Battle of Ypres. He'd read about the battle when a local boy's family repatriated him after dying in the Battle of Somme. Arthur Ackerman, a member of a prominent family, served at Ypres, too. The newspaper quoted him as saying, in response to a question about Ypres, that it was "*like being eye deep in hell.*"

Arthur's body arrived in Peterborough on November 8. Howard recalled the Fire Department being asked to take part in a large funeral procession that ran from the Armoury to Little Lake Cemetery. A local paper covered the story of the battle.

In the first week of April 1915, Canadian troops moved from their quiet sector to a bulge in the Allied line in front of the City of Ypres. This was the famed—or notorious—Ypres Salient, in Belgian Flanders, where the British and Allied line pushed into the German line in a concave bend. The Germans held the higher ground and fired into the Allied trenches from the north, the south and the east at will. On the Canadian right, two British divisions, and on their left a French division, the 45th (Algerian).

On April 24, 1915, the Germans struck, hoping to obliterate the Salient once and for all. A chlorine gas attack followed a violent artillery bombardment. The target was the Canadian line, and the Germans outnumbered them 5:1. Here, through terrible fighting, shredded by shrapnel and machine-gun fire, hampered by their issued Ross rifles which jammed; violently sick and gasping for air through soaked and muddy handkerchiefs, the Canadian Forces held on until reinforcements arrived. Their high spirits and courage, the only things that carried them through the ordeal.

General Manager Denham retreated towards a group of office staff huddled a safe distance from the fire. The first person he spotted was Eileen Donoghue. Denham sent her on a mission to recruit other girls to set up phone communications as fast as possible. He suggested Eileen go see Robert Munro, proprietor of the Munro House, a hotel at the corner of Hunter and Water Street. The hotel was an ideal place to establish a temporary base of operations, being so close to the plant.

Eileen found Amy Jamieson, and the pair hurried down the street. Munro was eager to be of help, considering that between the war and the Christmas Season, business was almost at a standstill in the hotel. The girls' dishevelled outfits, both muddied and tattered, must have been a sorry sight. Munro took pity, gave each a stipend and sent them over town to get new outfits.

While they freshened up, Munro had a telephone operating machine moved to the hotel so the girls could work. He had a phone line re-routed from an existing one 200 feet away. Telephone Company employees spliced it and ran a line into the hotel, dedicated to the operating machine.

There were too many among the injured and too little safe space around Dr. Birnie's car to triage there any longer. What seemed an eternity was in reality less than a half an hour. The local ambulances, both privately operated, arrived on scene as soon as news got to them. Men with farm wagons appeared out of thin air to offer help in transporting patients to the hospitals as well. The ambulances would never have been able to keep up on their own.

Despite the ramshackle look and a distinct odour of manure, the farm carriages were a welcome sight to those attempting to get the injured to hospital. They placed blankets as a barrier on the floorboards, then packaged and loaded patients. The ensemble headed east on Hunter Street in a macabre parade. Uninjured factory staff acted as orderlies and escorted the wagons to hospital.

Since St. Joseph's Hospital was the closest to the plant, the first wave of patients went there. Dr. McNulty ordered the most seriously burned and injured to be treated in the operating room. The nursing staff assembled equipment, blankets, bed linens, dressings and other supplies. They set up dressing tables on the main floor, in the corridors, and in the operating room. It didn't take long to be overwhelmed. There were men laying in every inch of space and more piling in through the doors by the minute. A stench of burned flesh permeated the building, along with pleadings for help as the nursing Sisters cared for the injured.

The scene looked more akin to a Medical and Surgical Hospital (MASH) unit on the front lines of war-torn Europe than a small-town medical facility. Semi-conscious men with ashen complexions lay everywhere, their clothing burned off and in tatters. A few screamed hysterically from the pain. The worst cases received morphine before being sent up to the operating room for treatment.

The City's doctors soon arrived in greater numbers and staff gained ground on the chaos within the building. Some victims came in without vital signs. A screening process prioritized the critical cases from those not as severely injured, and the dead separated from the dying. The most acute, sent to the operating room, and the rest treated where they lay. Nurses used every inch of corridor and unused space within the hospital. There were not enough beds in the hospital to accommodate so many new patients.

It became clear that any admitted in hospital before all this would have to be sent home, if able. Under the direction of Dr. McNulty, staff assessed those patients and discharged them. There were only 35 beds at St. Joseph's and space within the building was in high demand.

Orderlies placed blankets, hot-water bottles, and heated flat-irons around the wounded to combat shock. Doctors gave inhalations of smelling salts and offered whiskey in milk to those who could swallow. To help relieve the agonizing pain, medical staff administered Laudanum. There were many who had fractures to cope with, as well as burns. The staff of St. Joseph's Hospital put in a heroic effort to save as many as possible.

Within an hour, throngs of by-standers and those anxious about the fate of loved ones materialized in front of the factory. The menagerie was more riveting than anything the Ringling Brothers dreamt. Buckboards, coaches, and wagons lined up and down Hunter Street as far as the eye could see. The Police Officers, under Chief Daniel Thompson, struggled to keep the crowds at bay while the Fire Department undertook rescue operations.

"*Get these gawkers, back, damn it!*"

Chief Thompson barked the order to a group of soldiers he'd seconded to assist the police officers in crowd control. It was apparent the throngs of people who'd shown up to watch the factory burn had no sense of their own mortality. They were underfoot as emergency personnel worked, standing mere feet from the buildings.

The intensifying heat soon drove most of the people back, and they retreated to a safer distance, without further insistence from the soldiers. Several onlookers found a vantage point along Driscoll Terrace, on the opposite side of the Otonabee River. The fire transfixed any who witnessed its sheer power. Between the sight of the flames licking at everything within reach and the voluminous roar of the beast, it was oddly breathtaking.

Priests and clergy of many denominations arrived. Some stood and prayed, arms outstretched towards the heavens, seeking intercession from this horrific calamity for anyone beset by it. Others preferred to get into the thick of things and assisted Dr. Birnie in administering to the injured. Everyone moved forward with a common goal in mind, to help and serve others. This was also true of the everyday citizens of Peterborough, who showed up in numbers to help. They pulled together in a heart-beat to support those affected by this tragedy. Without hesitation, women arrived bearing food baskets for the firefighters and emergency responders. Men offered use of horses, vehicles, and even themselves where required.

Yet to Chief Howard, standing back and looking at the destruction, it seemed they were treading water. He realized everyone stood trapped in an hourglass full of sand. Buried in degrees, time running out, no escape.

Chapter Ten

A Trying Time for All

> *"Character cannot be developed in ease and quiet. Only through experience of trial and suffering can the soul be strengthened, vision cleared, ambition inspired, and success achieved."*
>
> - Helen Keller

Superintendent Ross Dobbin and the hard-working men of the Waterworks Division closed the supply valves at both ends of Hunter Street. The benefit was immediate. Pressure began rising in the underground system of pipes. Before long, the needle on the gauge for that end of town held confidently at 100 pounds per square inch.

The Fire Department noticed the change like night and day. Attack lines lunged forward in the hands of the firefighters, and water spewed forth under pressure to their great relief. Chief Howard ordered a team of the men to gain access to the Quaker hydrants on the west side of the building. A barrage of flames and intense heat drove them back.

Undeterred, they attempted to make hydrant connections farther up Hunter Street. The mission, to protect the homes lined along Sheridan Street, running north from Hunter Street just west of the factory. Their wood-frame construction and shingled roofs were no match for the tempest bearing down on them. Since the factory was a losing battle, the Fire Department regrouped to save the rest of Peterborough.

In the end, there were fifteen streams of water playing on the fire. They used between five- and ten-thousand feet of hose-line, twice as much as the Fire Department owned. But private companies like Quaker Oats kept a supply of hose, too. Chief Howard had requisitioned Quaker Oats' early in the emergency.

The Fire Department borrowed hose from the Henry Hope Company, too, and Bonner-Worth Mills. When Lindsay Fire Department arrived, they contributed a supply of their own.

Just before noon, staff at the Waterworks noticed the new water pumps installed at the pumping station had overheated. The culprit was inconsistent pressure for most of the morning. The pumps needed to cool before irreparable damage occurred.

Under normal pressure conditions, these pumps ran under heavy load all day long. But, when the pressure fluctuated as it had been, the pumps overheated. If not shut down, they'd fail, seizing from the heat. Dobbin realized it would both compromise firefighting activities, and the entire city left without water if the pumps failed.

Superintendent Dobbin notified Chief Howard and Mr. Denham, and ordered the pumps shut down for a half hour. It forced the firefighters to suspend suppression activities. They saw, of course, how it looked to the gathered spectators when they became bystanders themselves. Firefighter Frank Degan commented to the team,

"*Well boys, guess it's about time to unbutton our pants!*".

Their frustration was palpable. It didn't take long for the wolves to circle. The first to approach Fire Chief Howard, the County Clerk, a man named Edward Elliott. A few moments before, Elliott was having a heated conversation with Mayor JJ Duffus, and the other City and County staff members who had congregated at the fire. Elliott strode towards Chief Howard with purpose, the index finger of his right hand pointing upwards.

Mr. Elliott barely had time to catch his breath before taking Chief Howard to task over the clear lack of concern for the Courthouse building. Elliott wanted to know why nothing was being done to protect it from harm. Howard let him vent for a few minutes as the tirade continued, before cutting him off mid-sentence.

"*Mr. Elliott, I do not owe you or anybody here, including the illustrious Mayor Duffus, a play-by-play analysis of how I do my job.*"

Chief Howard glanced in the general direction of Mayor Duffus before locking eyes once more with the County Clerk, gauging the Mayor's reaction. The shocked countenance on Elliott's face screamed that the Chief had a lot of nerve speaking that way, but he pursed his lips and said nothing in reply.

"*However, since I have a moment to spare here, let me explain something to you, and I want you to listen carefully, because I am only going to say this once. Do you see those buildings over there, son?*"

Howard pointed towards the houses along Sheridan Street. He chose the words with care, for greatest effect. Elliott's eyes never wavered from the staring contest they were having. A defiant gaze bored a hole through the back of Chief Howard's skull.

"*Those houses are wood frame construction, built with highly combustible materials and having shingled roofs. They stand right beside the biggest bonfire that this city has ever seen. Left unattended, they will ignite, and like a row of dominoes, take the whole damn city with them. Is that making any sense?*"

The Chief asked rhetorically, and when Elliott opened his mouth as if planning to speak, Howard didn't give any quarter in which to answer.

"*The Courthouse up there, it's built of solid stone, has a metal roof, and is upwind of this fire. And I only have enough personnel and equipment to save one: either the entire City of Peterborough or the Courthouse. So why don't you tell me which you would pick if it was up to you. Take your time.*"

Once more, Howard gave Elliott no opportunity to reply, but turned on his heels and walked up to the Courthouse for his own peace of mind.

Constructed of solid stone quarried from nearby Jackson's Creek, tin covered the Courthouse roof. Howard felt confident the building was immune to the Quaker Oats fire, being upwind of the factory. After finding the building untouched, Howard headed back to the Quaker Plant.

No sooner had Howard arrived back, when he learned that the County Building, located beside the Courthouse, had caught fire. He didn't want to believe it. Howard directed a crew to lay a line of hose from a hydrant on Water Street and proceed up the steep hill to the County Building.

The line was too short, and the Fire Department was out of hose. Lindsay Fire had not yet arrived. The Chief tasked another crew to take the supply wagon over to the Armouries a few blocks away and borrow their hose. But when the firefighters reached the Armouries, the soldier in charge refused to permit the removal of the hose without written permission from the Lieutenant-Colonel.

The supply wagon returned to the scene empty-handed. Chief Howard was furious. But there were more pressing issues at hand. He told the crew to disconnect a section of hose-line on Hunter Street. Once done, Howard directed them to drag it up the hill on Sheridan Street, to the front of the Courthouse. The steep hill was glare ice from water travelling afield from the Quaker fire and coming to rest on the surface.

With a heavily laden wagon and the team of horses, the firefighters attempted multiple times to make the grade. Spectators jumped to the task, with some pushing from behind and others pulling from the front to assist the firefighters. Twice, one of the grey horses went to its knees before reaching the top of the ascent. If not for the passersby, the wagon may never have made it.

In the meantime, the Courthouse had lit up, and the fire gained headway with lightning speed. Unknown to anyone, a firebrand had found an exposed bit of fascia on the Courthouse roof that was unprotected, and seated itself. Lindsay's Fire Department arrived on scene by train as the local crews struggled with this new emergency and took over at the Courthouse.

It was now their turn to struggle to establish a water supply. Not only did the City's water pump still need to cool off, but Lindsay's hose connections presented an unforeseen challenge. The threads in the hose-couplings did not marry with the hydrants in Peterborough, which were a unique pattern. Firefighters had to cut the couplings off the ends of Lindsay's hose and rig their own unions onto the line to connect to hydrants. The delay cost them. Chief Howard sensed before this was over, he'd eat crow with County Clerk Elliott.

The lunch bell rang at St. Peter Elementary School, signalling the end of studies for the morning. Unlike King George School and Immaculate Conception, the students of St. Peter's did not get sent home early. While the school officials were aware of the situation at Quaker Oats, the school sat far enough away from danger. It was business as usual for the rest of the morning there.

The O'Brien children did not dawdle on their way home. Rumours of a large fire at the Quaker Oats factory circulated the schoolyard, with the same urgency as the flames consuming the plant. They spilled through the front door, falling over one another, a boisterous quartet asking questions Laura didn't know how to answer.

"*Momma, momma, is daddy ok?*" asked George as he tugged on the hem of her skirt.

"*What happened, momma?*" Joe asked, before Laura got to answer George.

"*What about grandpa Hogan?*" Irene blurted out apprehensively.

Laura had nothing to tell. She hoped to keep them calm through lunch and not alarm her children unnecessarily. So, she explained that she only knew there was a fire at the factory. While dishing beef soup into bowls, Laura diverted further questions by telling them to go wash up before lunch. She fought to keep her own composure, finding it harder than expected.

As the children devoured a modest meal, the conversation returned to what might have happened at their Daddy's work this morning. They told Laura how the teachers whispered and a few had even cried, and none of them seemed to know what happened. One talked over the other as the volume of their chatter rose. The children's precociousness was innocent enough, but Laura struggled to hold it together.

"*All right, children, all right. One at a time, please. Remember your manners? Goodness gracious!*"

She didn't want the children to see how she felt at that moment and used small talk to avoid the hard questions. Laura asked what they learned in class that morning and if they behaved for the nuns; whether they saw interesting things on the way home from school; and anything else to side-step the uncomfortable queries.

With a silent sigh of relief, Laura looked up at the clock and noticed it was time for the kids to head back to school. As she assisted them in donning mitts and toques, Laura kissed each one and said she loved them very much. As if on queue, the unborn infant growing in her womb moved. Laura started for a moment and rubbed her growing tummy.

As Laura waved goodbye to the four, hoping she had dodged a bullet, Kathleen turned back and asked,

"*Will daddy and grandpa be home when we get back from school, Momma?*"

Laura's heart caught in her throat, and she barely managed a whispered response in a failing voice.

"*I hope so, honey! See you after school.*"

She turned away before any of them saw her eyes fill with tears, gathering little Michael's hand, and headed back into the house.

Laura's father, William Hogan, stayed with the O'Brien's as he had done every December for several years. As a community minded man, William sought employment at Quaker Oats each Christmas Season. He used the money to supplement a meagre farm income and make donations to the Knights of Columbus' Christmas Toy Drive.

Hogan used little of the money for himself, but worked so that others had a better Christmas. While hard, life had been good to William and wife Annie. For a few weeks around Christmas, his job allowed them to help those less fortunate. He stayed with Laura and Dennis because it was closer to work, but also gave him more time to spend with the grandchildren.

At 72 years of age, William had retired from farming. When not in the employ of the Quaker Oats Company, he lived on a rural property outside of Peterborough in Smith Township. He and his wife Annie had raised a family there. Laura, christened Julia Lauretta, was the second youngest of nine children.

Laura prayed her father would soon be home from work so she could find out what had happened that morning at the factory. As she began getting his post-work snack prepared, the thought that her own Daddy could soon put her fears to rest once and for all comforted Laura.

Chapter Eleven

A Community United

> *"I daresay it seems foolish; perhaps all our earthly trials will appear foolish to us after a while; perhaps they seem so now to angels. But we are ourselves, you know, and this is now, not some time to come, a long, long way off. And we are not angels, to be comforted by seeing the ends for which everything is sent."*
>
> - Elizabeth Gaskell, Wives and Daughters

A veritable convoy of farm and delivery wagons, along with buggies assisted the two ambulances in town, operated by Daniel Bellegham and Aaron Comstock, to get the injured to hospital. In a matter of hours, they managed to triage and package everyone rescued for an uncomfortable ride to hospital. The ambulances alone would never have kept up without the help of by-standers, local farmers, and the citizens of Peterborough.

The student nurses' once smart white, wide starched smocks were a soiled, sodden mess. Their navy overcoats, as filthy as their winter boots. Still, they worked tirelessly with other nurses and the doctors. Until Bellegham, Comstock and the ragtag shuttles had transported the last injured person from the scene, the young ladies never wavered. After the worst of it, Dr. Birnie relieved the girls from duty and sent them marching back to the hospital. She remained behind in case anyone else needed medical aid.

At both hospitals, the sudden influx of patients exhausted available medical supplies. Sterile dressings and salves were especially scarce because of the sheer volume of burn victims. Staff prepared tannin solutions from steeped tea and began tearing up bed linens for makeshift bandages. They used sheets soaked in ice water to swaddle the most seriously burned and offer temporary relief, but the usual provisions soon became depleted.

Mr. Frank O'Conner received an emergency call late that morning from the Sisters of St. Joseph. O'Connor, a native of nearby Desoronto, had worked at the Canadian General Electric in Peterborough as a young man. He, along with wife Mary Ellen, founded the Laura Secord Chocolates Company. When he heard of the tragedy in Peterborough, O'Conner called St. Michael's Hospital in Toronto and requested a train car filled with medical supplies and shipped to Peterborough forthwith.

At around 1 pm, that train arrived at the Ashburnham train station. An army of volunteers hand-bombed the much-needed supplies up to St. Joseph's Hospital. They could not have come a minute sooner. Another thing in short supply was extra sets of hands. The influx of injured patients overwhelmed medical staff, and rest wasn't coming anytime soon.

Dr. Birnie let out a sigh of relief when she noticed Anishinaabe Elders and Healers of both Mud Lake (Curve Lake) and the Mississauga's of Rice Lake (Hiawatha) First Nation. They'd arrived out of nowhere to help. She welcomed their traditional healing methods as a practitioner of natural medicines. Dr. Birnie took a break from the hectic pace of the fire scene to acknowledge them. Covered from head to toe in soot and blood, she was an unsightly mess, striding over to their buckboard wagon.

"*Aaniin and miigwech.*" she said in Ojibwe, meaning "*Hello and thanks.*"

She'd picked up a smattering of Ojibwe while learning Indigenous healing methods and knew many of the Elders present. Dr. Birnie asked if they'd perform a smudging ceremony. Such a rite took away negative energies. Others offered prayers to the Creator for the injured and invited Spirit Guides to attend those who had died. Dr. Birnie spoke to the Healers next.

From the time humankind harnessed the power of fire, they needed treatment for burns, scalds, and blistering. Indigenous People learned from nature which remedies worked best. Long before the common use of medicinal creams, or antibiotics, they learned which plants produced antioxidant, anti-inflammatory, and antimicrobial effects.

Those First Peoples identified over 400 different species of flora with medicinal applications. Dr. Birnie remembered them calling these plants, "*nanaandawashkwe*", meaning sacred herbs. Honey, too, being made of various sugars is renowned for its highly viscous properties. It absorbs fluid from wounds and pores and allows healing to occur in a moist environment. Honey possesses not only antibacterial properties but anti-inflammatory and anti-fungal ones, too.

In the language of medicine, Dr. Birnie might say it helped stimulate the growth of tissues and reduces edema, inflammation, and pain. And she observed that the concentration of honey is proportional to its antibacterial benefit. The higher the concentration, the better the antibacterial properties. Honey showed better healing rates compared to other forms of pharmaceutical treatments.

Dr. Birnie learned from the local Healers how to make plant-based topical treatments using plantain leaves or dandelions. They taught her that ingesting red clover flowers had therapeutic uses and anti-inflammatory and antibacterial effects. Applied as a poultice on the skin, the roots of cattail plants helped heal burns and skin infections. The pitch from spruce mixed with grease from animal fats helped to heal burns.

The Healers who arrived to the Quaker Oats by early afternoon brought prepared tinctures and poultices to soothe burns for infection control. Honey was in their arsenal too, both as a topical antibacterial ointment and to decrease skin contraction from the burns. Their noble offer of help was of great help to the local medical staff.

Other medical volunteers and even local veterinarians assisted the primary medical teams. Armed Forces veterans acted as orderlies and security. Before long, the family members of Quaker employees appeared unannounced at the doors of the hospitals. They were looking for loved ones and asking to see them.

Most were frightened, and a few became impatient with staff. They couldn't comprehend the chaos of the scene or why the admitting clerks didn't have a list of patients at their disposal. Overcome with anxiety, no one grasped why a doctor wasn't available to speak to them on the condition of a husband or brother. They didn't understand why they weren't able to see their loved ones.

As they shouted over one another and jostled for position, the mob of local citizens had become unruly in the front lobby of St. Joe's Hospital. Their worry escalated, patience wore thin, and numbers grew by the minute. The Armed Forces veterans were a thankful support to hospital staff and the local constabulary.

In the bowels of the building, the morgue was filling up, and the staff there needed a break. Several patients died in hospital. One of those was Richard Healey. He came in with serious injuries. The worst, that one side of his skull had caved in. He sustained the injury when the north wall of the Dry House cascaded through the roof of the Boiler Room.

Healey's vital signs were weak and thready when brought in to hospital. Despite their best efforts to revive him, Healey became vital signs absent soon thereafter. Dr. McNulty pronounced it and called the time. Orderlies removed Healey to the morgue. As the full-time morgue attendant left for a break, he put responsibility for the morgue with a local veterinarian who had shown up to offer help.

A few minutes later, investigating the sound of gurgling, the veterinarian almost fainted when he found Richard Healey, struggling to speak. Healey had peeled the dressing sheet away from his face. The nervous animal doctor called for help and raced the gurney back upstairs to the operating room, which had become ground zero for the most life threatened. The doctors worked to keep Richard alive a second time, but his life was fading away once more. Dr. McNulty brought him back a third time, and after stabilizing Healey, returned him to the wards.

People rallied to support those who needed help. Volunteer firefighters from the old Brigade and the former Ashburnham Fire Department arrived to help Chief Howard and his small crew. Soldiers who had returned from the war offered to assist the Police Department. They performed crowd control and security and acted as orderlies at the hospital. Anyone with any medical training relieved the overwhelmed doctors and nurses, including the Indigenous Healers.

Many who called the Peterborough Area home considered it an obligation to do whatever they could in this time of need. There wasn't a single neighbourhood unaffected by this tragedy. In that spirit, women such as Mrs. Mina Rogers rallied to help the families whose lives the fire altered forever.

Mina Rogers was but one of many women who, as part of both the Anglican and Methodist Church groups, organized an annual Christmas Food Hamper Program. In other years, Rogers and her team bought, packaged, and delivered food to the less fortunate of the population, regardless of religious affiliation. After hearing of the horrible loss at the Quaker Oats Plant that morning, Rogers contacted members of the Food Hamper committee.

Within hours, a plan was in place to speed up the Hamper Program. The team filled and delivered baskets of goodwill to any Quaker Oats families affected by the tragedy. Rogers knew that loss of life or debilitating injuries were dreadful any time, but this close to Christmas it was devastating. Many Peterborough families were living paycheque-to-paycheque. So, the parishioners of various church communities stepped up to support their community.

There was a multitude of smaller, unselfish and voluntary offerings made, too. Case in point, Mrs. Rebecca Hartley. She lived in a house at the end of Murray St, near the plant. Rebecca kept a wash-boiler filled with coffee for the firefighters throughout the day.

"*Quaker Oats Company, how may I direct your call?*"

The kind and even voice that answered Denham's call belonged to one of the many hard-working switchboard operators at the Head Office in Chicago, Illinois.

"*This is William Denham, General Manager of the Peterborough Operations. I need to speak with Robert Stuart Senior, please. It's urgent.*"

"*Of course, Mr. Denham. Please hold.*"

William Denham waited for what felt an eternity. His right leg bounced up and down in rapid succession until he noticed the desk shaking and stopped.

Nerves, he thought, and said, "*hold it together, Bill, for God's sake.*"

Robert Stuart would not be happy to hear the news. Even then, from the relative safety of the Munro House, Denham heard the muted explosions, breaking glass and deep, slow, angry groaning of large structural beams inside the factory. Perspiration formed in beads on his lower lip. The back of his mouth resembled the Sahara at noon, dry and gritty. As Denham motioned to Robert Munro to bring a glass of water, the series of distinct clicks on the line snapped him out of the darkness of his thoughts.

"*Mr. Denham? I have Mr. Stuart on the line for you now. Please go ahead.*"

"*Thank you. Ah, Mr. Stuart? It's William Denham here. From Peterborough. I have bad news, Sir.*"

But to Denham's great relief, company owner Robert Stuart wasn't angry. In fact, Stuart promised that he and a team of representatives from Head Office would leave as soon as possible for Peterborough. Denham's placed the call around 11 am and within two hours a group of Quaker Executives, including the ownership team, boarded a train in Chicago.

In that very brief time, Quaker coordinated with multiple railway companies to clear the tracks. The train needed to travel unobstructed from Chicago to Peterborough, without the need to change trains anywhere along the route. A massive coordination effort pulled this off in just two hours. They placed phone calls and/or telegrams to not just one, but multiple transit authorities managing train track systems spanning International borders.

From Chicago, company representatives secured, supplied, and staffed the train and engine. They arranged accommodation and transportation for when in Peterborough; and negotiated with governmental authorities of two countries to expedite crossing the border into Canada. It is a testament to how committed the Quaker Oats Company was to their staff, no matter where in the world. By 1 pm, they were on a train headed north. A little over twelve hours later, the group arrived in Peterborough.

The well-seated fire had consumed much of the north end of the factory by early afternoon. As walls, floors and the roof continued to collapse, it threw fire brands up into a swirling wind. Citizens and business owners alike looked for any means necessary to prevent the loss of property.

Many resorted to watering their homes and buildings with garden hoses to prevent ignition. They stood guard against fire, despite the frigid temperature and wind. But other buildings did not fare well under the circumstances. Not only had the County Building and Courthouse caught fire, but a lumber yard across the river was soon ablaze. The Fire Department had to abandon the Quaker property to knock that fire down.

They brought the lumber yard blaze under control within an hour. No sooner did they arrive back to Quaker Oats, when more chemical tanks in the factory started exploding. Loud detonations sent onlookers scrambling back in horror. Great jets of flames shot straight up into the air above the mill and the noise was frightening. This further deterioration of already hostile conditions tested the resolve of even the most hardened firefighters.

The fire had a fresh source of fuel and hungrily moved southward. It ignited the last vestige of the factory complex, the adjoining administrative offices at the south end of the property. Flames had reached Hunter Street unabated. As flying hot embers inched ever closer to downtown, another threat emerged.

The City's Gas Works, located a couple blocks away, near the corner of Simcoe and Queen Street, was in danger of igniting. Chief Howard realized if the fire reached the gasworks, the explosions at the Quaker factory would seem as tiny pop rockets in comparison. He dispatched a crew to assess the circumstances downtown and recommend evasive action.

There were no formal evacuation plans for the City. No one envisioned an emergency of this magnitude. Despite that, most residents on Sheridan Street had already bailed out. A few stayed to direct small streams of water on their houses with garden hose. But if the gasworks blew up, the loss of life and property would be substantial. Except there was no straightforward way to warn the inhabitants of downtown.

By 4:45 pm, the upper floors and walls of the Dry House and both Warehouse buildings had fallen in on themselves. It was a blessing for the fire suppression crews. As the buildings collapsed, it snuffed out large sections of the fire and cut off its available air supply. Unfortunately, it made the job of the rescue crews difficult when the time came to locate bodies.

The inferno was still hemorrhaging at the south end of the building, though, even well beyond the 6 pm hour. If fire crews took a positive, the bonfire illuminated the scene. The sun had set and plunged the City into darkness, shortly after 5 pm on the wintery December evening. To supplement the glow of the flames, they brought in and set up electric lighting around the building. The Fire Department did not plan to stop for the night.

Earlier in the day, as the ladies of various churches organized a food drive, another group of residents planned a prayer vigil. The service was at the factory grounds at 7:15 pm. Armed with candles to dispel the darkness and defying of their growing fear, a throng of brave people stood in silent reflection. As the vigil started, clergy led petitions for the fallen and injured.

The fire raged in mocking indignation, mere feet away. It was not only alive, but full of anger and violence. Indeed, it threatened to consume the very ground upon which the group stood. The assembly replied, offered prayers of peace. As the firefighters toiled vigorously around them, the faithful dispersed in silence. It was inappropriate to discuss the loss of so many this close to the assassin.

Well-wishers, neighbours, and family friends descended upon the home of Dennis and Laura O'Brien, beginning around the dinner hour. The kids sat around the kitchen table when the first knock came to the door. Laura wasn't expecting company and by then, beside herself with worry because neither her father nor husband returned home. Two men Laura didn't recognize were standing on the front stoop. They were not even 20 years old, if she guessed.

"*Mrs. O'Brien? Is Dennis home?*"

Laura's faced flushed as she rang her hands together. That the men assumed Dennis was home brought the worry to a fevered pitch. Her voice shook as she replied.

"*I'm sorry, he isn't. Is there something I can help you gentlemen with, instead?*"

The two exchanged confused glances before one looked at her, hat in hand, and spoke. Laura could see what looked like dirt covering both men in the evening's dusk.

"*Well, we just dropped by to thank Dennis for saving our lives today, ma'am. We wouldn't be alive if it weren't for him.*"

Both were uncomfortable. The man speaking rocked back and forth, shifting his weight from one foot to the other, and looked down at dirty, worn shoes. Laura tried to process what he said, unsure how to respond, before finding the courage to speak.

"*The truth of it is, Dennis didn't come home from work today. And neither did my father, William Hogan. What happened at work today?*"

She gathered from what John Todd said, there'd been an explosion. Laura felt it rock the house. What she wanted to know, where her husband was.

"*Well,*" said the second man, "*Dennis was a hero today, is all. He sacrificed his own safety to rescue the both of us from the clutches of the flames, Mrs. O'Brien.*"

Laura had to face the grim reality that something went horribly wrong. At first, she considered getting a neighbour to watch the kids and walking over to the factory. But, as people continued to flood into her home unannounced, Laura realized that was out of the question.

News travelled quickly that Dennis and William didn't return home. Neighbours, friends, and even family came to their Harvey Street home to support Laura. Except she didn't even know if she needed support from them. She was in shock, but still hoped Dennis and her father were still alive and working with emergency crews. In Laura's mind, it was not time to plan a funeral just yet. Still, there were many acting as if that was a foregone conclusion, dropping by with food and condolences.

And then the reporter arrived. As he introduced himself, Laura felt her knees buckle before everything went dark. Her last thought before losing consciousness was that reporters don't show up at your house unless something bad happened.

Chapter Twelve

Into the Aftermath

"Like after a prairie fire… It seems like the end of the world.
The earth is all scorched and black and everything green is gone.
But after the burning, the soil is richer, and new things can grow….
People are like that, too, you know. They start over.
They find a way."

- Celeste Ng, Little Fires Everywhere

The *Peterborough Examiner* front-page article of Tuesday, December 12, 1916 said it best in terms of the scene at Quaker Oats on the next morning:

"A smoking heap of ruins, outlined by a jagged edge of wall, is all that remains of the front part of the main building of the Quaker Oats Plant. A trip to the scene of the catastrophe this morning reveals little change from last night. A part of the fireproof warehouse on the east side of what was once the main plant, is still standing, while the rear of the main building and close to the scene of many miraculous escapes stands a portion of the brick elevator. Farther to the rear and standing alone amid the debris is a battery of sixteen concrete silos. With the last named exception, the entire plant is a complete wreck."

Few associated with the fire got more than a wink of sleep Monday night. Fire crews worked through the night to surround and drown the flames in place. Staff at the hospital put in overtime to care for the injured. Denham was beside himself with worry. Phone calls and meetings occurred until well after midnight. Mayor JJ Duffus gave up a night's rest to be at the train station to meet the Quaker Executives at 2 am when they arrived.

As trying as it was for the emergency personnel and first responders, others put in a sleepless night full of anguish and grief. Dennis O'Brien and William Hogan were not the only men who didn't return home from work on the evening of December 11. Jack Conway, Albert Staunton, George Vosbourgh, William Mesley, James Gordon, James Foster, Alphonse McGee, Richard Chowen, and Wilbert Kemp were among those reported as missing. For their families, who felt helpless, the wait was agonizing.

Not knowing the whereabouts of a husband, father, brother, or son was the worst likely scenario for those families. It provided no closure. A close second, learning a loved one died. Those recovered dead on scene or who succumbed in hospital on Monday afternoon included Vincenzo Fornaro, Domenic Martino, Filippo Carbone, Ned Howley, Thomas Parsons, and Walter Holden.

The injured admitted into one of the two local hospitals included John Kemp, Patrick O'Connell, Michael Long, James Murphy, Edward Bedding, William Garvey, Richard Healey, William Teatro, William Walsh, Joseph Houlihan, Mauro Giardini, John Weir, and Miss Eva Booth. Many sustained horrific burns, and a few, not expected to live. A hard pill to swallow for their families. But at least a slim hope of recovery glimmered on the horizon.

Over breakfast that morning, Laura struggled to explain to the children what happened. Because of their age, they didn't fully comprehend what she said. A few remained optimistic that their dad and grandpa were alive, and George planned to prove it.

After school on Tuesday, December 12, he escaped the watchful eye of older brother Joe and made a bee-line straight to the factory. As his siblings turned north on Water St. from Brock, George O'Brien headed south towards Hunter. They didn't notice him missing until arriving home.

"*Where's George?*" Laura asked Joe. She was cross with Joe and spoke in a shrill tone, scolding him.

It hadn't been a good day. As George's siblings stared into mittened hands, unsure of how much trouble they were in, Laura continued.

"*Joe, go find your brother. Now.*"

"*But…*"

"*No 'but's', Joseph O'Brien. Get! He's likely gone to the factory. Start there.*"

Joe planned to mount a protest, but Laura used his full name. The surest sign of trouble was when a mother resorted to such tactics. Joe mumbled a defeated '*yes, ma'am*' and turned to walk away. The others stood frozen in place, not wanting to draw her impatience, either.

"*Joe.*" Laura said as he turned. "*I love you. I'm sorry I yelled, son. It's been a hard day. Do you understand?*"

Joe nodded a yes. He continued south towards Quaker Oats. It wasn't the sting of her tone that hurt. He was mad at himself for not noticing that George disappeared. Even at the tender age of ten, Joe realized he became the man of the house if anything happened to his daddy.

At the factory grounds, George explored a series of tunnels that ran through the basement of the plant. A sharp odour of burnt oats remained overpowering. The crisp echo of distant voices in the cavernous space provided a soundtrack to the search. Farther away, a man barked an order to someone. A distant thunder of scraping concrete and shriek of metal reached his ears, and George winced. Another man shouted. In the dim light, George tripped over debris and fell, scraping a hand enough to peel back the skin. Like any boy of seven, he washed the blood off in the ankle-deep pool of filthy water at his feet.

George thought, if his father suffered a head injury, he might have amnesia and unable to find the way home. In George's mind, Dennis and William were unconscious or trapped and waiting to be rescued; and he wanted to be the hero of the day. But unable to locate them in the dark, thought they joined the rescue operations. The light of day made him wince, emerging from the darkness of the tunnel.

"*What are you doing in there, young man?*"

George jumped, startled, and peered from beneath a cupped hand on his forehead to see the police officer.

"*I'm lookin' for my dad and grandpa. Have you seen them?*"

"*You're not allowed to be here. It's dangerous. You'll have to go home.*"

"*But my dad didn't come home from work yesterday. Can you help me find him?*"

The officer shifted on his feet, softened by the boy. He regarded George with a hint of amusement. Covered in mud and in soaking wet clothing, George's shoes made a squishy sound as he stepped closer to the officer. With each step, more water squirted out.

"*What's your name, son?*"

"*George.*"

"*George what?*"

"*George O'Brien, sir.*"

The officer was about to ask George if he lived nearby when a shout came from behind.

"*George!*"

Joe ran up, out of breath, and grabbed him by the arm. George pulled away, getting out of the grip, and stuck his tongue out at Joe. The officer stifled a chuckle.

"*Mom's worried sick. Here, don't you sass me. Dad's gonna tan your hide if you don't listen to me.*"

George stopped struggling against Joe's grip.

"*Is dad home?*" he asked, hopeful.

"*No, but mom sent me here to fetch you. Come. On!*"

Whatever thought took George to the factory, the steadfast belief that he'd see his father again compelled George to visit the factory daily in the coming week. Many times, he spoke to a police officer or firefighter to ask about his dad, and a few times got underfoot and told to go home. Often, Joe retrieved him, the scene from Tuesday replaying anew.

Laura O'Brien became far more realistic as time passed. Rescuers found her father, dead in the wreckage of the Boiler House. But Dennis' body had not been located. In the coming days, Laura heard many stories of Dennis' heroism. As proud as she was, Laura experienced moments of regret and wished he hadn't been such a hero and just come home to his family.

When the afternoon edition of the *Peterborough Review* came out, a large advertisement requested available male Quaker employees to report for duty at 6 pm that evening. Search and rescue operations would begin straightaway. William Denham placed Thomas J. Grant, a foreman in the plant, in charge of this rescue operation.

Mr. Grant, along with City Engineers, surveyed the damage first thing in the morning to make sure it was safe to be searching in the rubble. They decided an imminent danger of the walls collapsing remained. Grant ordered heavy machinery brought in to tear down barriers and stabilize the building so rescue operations could begin that evening.

After more of a long nap than a night's sleep, the Quaker Executives, including Robert Stuart, Sr. and son Robert Jr. headed to the scene of devastation that was the emaciated remains of the factory. Mayor Duffus, the Aldermen, and City officials including Fire Chief Howard gave a first-hand tour of the remains.

The Stuarts had many questions, of course. One of the first, how a Mill with a host of state-of-the-art fire safety equipment installed could lie in ruins in less than 24 hours. Neither Howard, Dobbin, nor City Engineer Roy Parsons gave any conclusive answers, and said it was their top priority to find out what happened.

For the moment, a mountain of bricks that glazed from the intense heat held their secrets. Covered in inches of frozen water, they concealed the severed water mains coming into the plant; the fractured sprinkler risers and fused sprinkler heads. No one knew yet that the blast warped fire doors and tore them off hinges, the splintered remains hidden among the debris.

Ross Dobbin only said that something caused pressure to plummet in the city's water mains. The inconsistent pressure prevented firefighters initiating an interior attack until the fire had overtaken them, a span of less than one hour. Chief Howard explained to Quaker management that it caught the Fire Department flat-footed and they were on the defensive from the beginning.

No one was assigning blame at this point. The Quaker brass just wanted to know what happened. Chief Howard, exhibiting a great deal of self-control, opted not to throw City Council under the bus. It would have been easy for him to justify that he begged for better apparatus and requisitioned more equipment, particularly the motor-driven pump. Howard didn't mention the department was thousands of feet short on hose, that 13 of the 19 ladders lay destroyed, or that said ladders could only reach the third floor of a building.

As they took a walking tour of the exterior of the property, surveying the destruction, the next question was to enquire about casualties. No one could give a firm answer. Newspaper reports were conflicting, and chaos reigned at the two hospitals. The consensus among the local group was the number of deaths could be as high as sixteen with scores injured.

Since time was of the essence, Mr. Stuart suggested they strike a team to attend both hospitals and interview survivors. This was important for any investigation by outside authorities. But as hands-on owners, the Stuarts also wanted to know first hand. Despite Quaker Oats being a multi-national company, the owners still took an active role in operations. They wanted to aid with an investigation and sort this out to prevent it from happening again.

One item on the minds of the local contingent was whether Quaker Oats planned to rebuild the factory in Peterborough. Robert Stuart assured Mayor Duffus and the Aldermen present that Quaker Oats had every intention of rebuilding the plant. In the meantime, there were more pressing matters that needed attention.

High on that list was getting operations up and running again. There were still wartime contracts to fulfil. Since that couldn't happen on site, the Quaker Executive team quizzed Duffus on the likelihood of renting mill space at other factories in town. Several smaller companies in the Peterborough Area would welcome the additional work, and Mayor Duffus promised to look into the matter and get back to Mr. Stuart.

Two locations came to mind, the Campbell Flour Mill and the Peterborough Cereal Company. The Stuarts and the Campbells shared a similar Scottish heritage, and Duffus had in mind to speak to Daniel Holland, the local manager for the Campbell Flour Mill. John Meyers was President of Peterborough Cereal Company. In a perfect world, both had capacity available to sub out some of Quaker's contracts.

As the initial site visit concluded, Stuart asked Denham to find out about the potential for increases in production capacity at other Quaker facilities in North America. By retooling manufacturing lines at other company-owned factories, they might avoid any sizeable gaps between supply and demand. This temporary measure provided a stop-gap until either they built another factory, or the war declared over. Also, moving the Peterborough staff to other Quaker properties meant fewer layoffs.

The group retired to a more comfortable environment for a hot lunch to continue discussions. As steam lazily hovered above white ceramic cups of black coffee, the Stuarts asked Denham for immediate next steps. He explained that two teams of Quaker employees would undertake a search and rescue operation that evening. These teams worked around the clock to account for as many of the missing as possible.

The conversation then steered to everyone's thoughts on how best to handle payroll. This close to Christmas was a trying time for the five-hundred Peterborough employees to be without work or income. Without hesitation, Robert Stuart stated that the Quaker Oats Company refused to see its most valuable assets in the lurch through no fault of their own.

The management team unanimously decided staff should receive one-month's wages. It bought time to decide the best course of action on production operations. They agreed to work with the newly formed *Workman's Compensation Board* to make sure the widows and any employees unable to return to work because of a permanent disability received an income.

Fire Chief Howard broke from the group when they set off for lunch, staying at the fire scene to be with his men. Deputy Chief Jamieson located Howard after rejoining the squad to inform Howard a telegram arrived at the fire station from Edwin Heaton, the Provincial Fire Marshal for Ontario.

Heaton's telegram informed Chief Howard that he would attend that afternoon to begin an investigation into the tragedy. A standard practice, Chief Howard looked forward to Heaton's visit. Not having been on the job long, Fire Marshal Heaton already had a reputation for being thorough and therefore an asset on scene.

Heaton arrived without fanfare or pomp. Edwin Percival Heaton was a no-nonsense sort of character who took the job seriously. Known as a perfectionist, Heaton bore little patience for half-measures. He was not in Peterborough for the "five-cent tour" of the destruction. Instead, Fire Marshal Heaton began the arduous task of determining what happened, deduce any wrongdoing, and to learn how to prevent a similar future tragedy from occurring.

People said his demeanour could be a "put-off" to others, especially towards those Heaton viewed as beneath him, or sprouting uninformed opinions in place of facts. Heaton did not have the patience for arguing, preferring to state things as he saw them, leaving no room for discussion.

As Her Majesty's duly appointed representative, this was Fire Marshal Heaton's scene, and he didn't mind telling that to anyone who begged to differ. He gave respect where due, and extended professional courtesy to Chief Howard, a man he regarded as close to equal under the circumstances. The two led the way together as the team surveyed the wreckage.

The building was a rats-nest of carnage. A mangled metal shell, heaps of shattered concrete, and the charred remains of inventory, the last vestiges of it. The sound of dripping water echoed through the barren structure. It broadcast within the factory like a ghostly message tapped out in morse code. The voices of firefighters reverberating off the walls deadened it. A pungent odour of scorched grain mingled with that of melted steel. Visibility was poor, even in the light of midday. Smoke shrouded their view as Fire Marshal Heaton led Chief Howard and Deputy Chief Jamieson through the plant.

Armed with lanterns, they surveyed the damage. One obvious truth was that the fire burned hot. The group was not far into the inspection when Heaton noted how the bricks from the walls had glazed and fused. He knew for that to occur, the fire had achieved a sustained heat of at least 2,500 degrees. Heaton further commented on how the concrete lintels of the windowsills had melted, evidenced by rivulets of what had been molten concrete oozing down the walls beneath the windows and for a distance of about eighteen inches across the floor.

Heaton pointed out that most of the floors sagged, stating it was obvious they'd been overloaded. He expressed surprise that a few even remained horizontal, other than the obvious sagging. As Heaton passed by one of the remaining concrete support columns, he took out a metal pick. He hammered on the outer surface, noticing the concrete had calcinated and how it powdered under the pick's blows. It further strengthened his theory that the heat attained as the fire raged was intense.

Such a volume of fire spread through the plant that exceptionally high temperatures developed. Because of the loss of the fire separations, the blaze did not stay contained, but spread unabated throughout the building. In places, entire wall sections had melted brickwork, sometimes to a depth of five inches. In others, only the metal shells of support beams remained; the thick concrete shell had burned and flaked away.

Outside the building once more, Fire Marshal Heaton directed the group's attention to a series of railway boxcars stationary on various sidings. The only thing left was the blackened skeletal remains of the metal boxcars. Even the chilled iron wheels melted, spilling down towards the rails like grey taffy. In one case, a wheel and the rail upon which it stood had fused together, the wheel indistinguishable from the rail. It didn't take a wizard to notice the fire burned with a ferocity rarely seen.

Mr. Heaton still had a lot to consider but announced he would lead a formal proceeding into the fire, bring witnesses and dissect the physical evidence. Until then, he had a lot of ground to cover. Before departing for the day, Heaton instructed Chief Howard to have barbed wire strung, signs posted, and guards placed to preserve the evidence. Until determining a cause, Fire Marshal Heaton considered the property a crime scene.

As afternoon wore into the early evening of December 12, the sun disappearing over the western horizon. Men in the employ of Quaker Oats began turning out as requested to begin a search and rescue operation. After assigning each person a complete rubber suit, including boots, hat and a coat, Thomas Grant divided the group into two gangs. The first began working at the north end of the plant, and the other at the northeast corner.

High-wattage electric lights, strung up on wires and draped over parts of the ruins, facilitated working throughout the night. It was punishing manual labour, operating out of doors in the cold, damp night air. Snow fell the previous night, coating the ground with a blanket of flakes. The fire consumed most of the plant, but deep-seated fires still burned in the peat of processed oats and Vim Feed, in the bowels of the building. It would be days before the Fire Department declared the fire out.

Unlike the raging inferno of the day before, the fire didn't give off enough heat to keep the searchers working outside warm. Their hands went numb from the cold, their eyes strained from toiling under less than optimal lighting conditions. To make matters worse, they worked on uneven footing. The trepidation felt was palpable. None wanted to find someone they had been working with mere hours ago dead in the rubble.

As rescuers, they had to put such thoughts aside and concentrate on the hope of finding a friend alive. In that spirit, the rescue teams worked hard all night. The work was punishing and done by hand. The only way they got through it was not to look at the utter destruction, but to focus on moving just one piece of concrete at a time. There would be plenty of time to mourn when they finished the job.

Chapter Thirteen

Of Casualties & Heroes

> *"That what I need to survive is not Gale's fire, kindled with rage and hatred. I have plenty of fire myself. What I need is the dandelion in the spring. The bright yellow that means rebirth instead of destruction. The promise that life can go on, no matter how bad our losses. That it can be good again."*
>
> - Suzanne Collins, Mockingjay

The search team worked tirelessly in shifts over the coming days. The grim task became more difficult with each body uncovered. Inside the remains of the Boiler House area, the team found William Hogan, James Gordon, Thomas Parsons, William Mesley and James Foster. Their deaths were swift when the north wall of the Dry House building crashed through the roof above them. The bodies had no burns, only the horrific injuries from being compacted in place.

Outside, on the grounds of the factory, the recovery team found the bodies of William Walsh, Ned Howley, William Miles and George Vosbourgh. Vosbourgh went unidentified at first. No one recognized him. It had nothing to do with injuries. The body was unburned, and face recognizable. That was the issue: he was clean shaven. George had sported a full beard for years.

When Mr. Vosbourgh gained employment at Quaker Oats a few months prior, they told him to shave off the whiskers. It was a Health & Safety requirement of employment in such an industry to be clean-shaven. Despite having reservations, George agreed to remove the beard. Even close friends didn't recognize him. Vosbourgh's was the last body recovered, found buried under snow and ice long after the rescue crews had disbanded. A farmer clearing snow near the Boiler House Area found him as the demolition team was getting ready to take apart that section of debris.

At the two local hospitals, the list of the injured shrunk as the list of the dead grew. Many succumbed to their injuries in those first few days.

The Fire Department did not declare the fire as under control until December 15, four days after it started. In reality, burning continued deep in the bowels of the factory for weeks afterwards. Because of the sheer volume of debris, crews couldn't get to the seat. Chief Howard called in Toronto Fire Services to help and provide more equipment, since flames destroyed much of the local Department's gear.

Meanwhile, news spread beyond Peterborough with the same ferocity as the fire itself. As far away as Australia, media reported on the tragic event. It put Peterborough on the map, but not in a way anyone wanted.

Officials took statements from the injured and dying in hospital first. Then, from the many who were fortunate to escape unscathed. Fire Marshal Heaton led proceedings that began on December 20, 1916, and Crown Attorney George Hatton examined the witnesses.

"*I want to make this very clear. This is not a persecution, or, a prosecution. It's not even a trial.*"

In that opening statement, Heaton clarified it was not an inquest. He knew there had been no foul-play, what transpired was merely an unfortunate accident. The investigation was a simple, honest effort to the facts concerning the disaster. To make sense out of the tragedy and to learn how to prevent such an occurrence happening again, he knew it was important to dissect the incident thoroughly.

"Chief Howard, others say that first explosion may have come from the boilers, in the Boiler House. What can you tell us, sir?"

Heaton knew the answer, but wanted it recorded in the official transcript of the proceedings, so directed the Crown Attorney to ask.

"I discovered the boilers intact, when crews excavated that area of the factory. And, the victims found there were unburned. The explosions didn't originate in the Boiler House, Mr. Hatton."

"Might the explosion have occurred in the ground hull tanks installed near the top of the building then, Chief Howard? The elevators led from them to the attrition grinders below, did they not?"

Heaton added that question too, wanting to end the rumours flying about. Since he wasn't providing testimony to the proceedings, Heaton knew it had to come from Howard.

"True. The grinders connected to the ground hull tanks. But as others testified, the tanks remained standing well after the explosion. If it happened in those tanks, fire would have spread across the adjoining bridge, into the Cleaning Mill. It didn't."

"Thank you, sir. And one last question. Why did the Courthouse sustain damage?"

Howard knew the question was coming. He looked around the Courtroom, and sure enough, Edward Elliott stood in the back, smirking. They made eye contact, and Howard never broke the stare as he answered the question.

"The roof of the auditorium of the Courthouse was lost because of the shortage of hose. And lack of manpower. Plain and simple."

General Manager William Denham, Waterworks Superintendent Ross Dobbin, stock checker John Cunningham, mechanic James Packenham and even eyewitness Annie Hopcroft from Driscoll Terrace provided testimony, too. Also read into the minutes were bedside statements provided by employees such as William Walsh and Richard Chowen, before they passed away.

Martin Kirn also provided testimony to the proceedings. As a chemical engineer with Quaker Oats, Fred Mesley and Robert Allen reported to him. These men were responsible to ensure the life safety equipment in the building worked at all times. Fred, during day shift and Robert, night shift. Fred was William Mesley's brother.

"*Mr. Kirn, please the Court, it's my understanding that you maintained the firefighting equipment in the factory. Is that correct, sir? What can you tell us about the condition of said equipment on the morning of December 11, 1916?*"

"*Yes, and I believe that all the apparatus for firefighting was in good condition at the time of the fire.*"

The Fire Department, along with Heaton, were not the only ones to investigate. As local crews excavated the remains of the factory, two other men arrived in Peterborough. T.D. Mylrea, and A.J. Mylrea were inspectors with the *British Fire Prevention Committee* and the *Canadian Fire Underwriters Commission*. The men were expert in fire cause and determination.

Both organizations published detailed reports. The well-respected independent agencies had studied every major incident involving fire within the Commonwealth. Their review encapsulated Fire Marshal Edwin Heaton's survey of the scene, the testimonies at the proceedings, and information provided from Quaker Oats Company management. They outlined what had caused the catastrophe, the worst in the City's history.

Long before the reports saw the light of day, however, a new firestorm was brewing at City Hall. There were those within City Council who were openly critical of Chief Howard's actions and advocating for his dismissal. It wasn't clear what they thought Howard might have done differently. But Council believed the factory would still stand if the Fire Department did their job better under the Chief's direction.

"*The man has to go.*"

Edward Elliott was, of course, referring to Fire Chief William Howard. He, along with Mayor Duffus and Alderman Johnson, were having a private conversation in the quietness of Council chambers. Elliott soldiered on.

"Alderman Johnson, you see that, don't you?"

As Johnson set the glass of Scotch on a nearby table to reply, Duffus interjected.

"Hang on a minute, here, Edward. What do you think Chief Howard might have done differently?"

"We almost lost the Courthouse and County Building. Thank goodness Lindsay showed up when they did. Did you see how that scoundrel talked to me? I told you he was trouble. Now maybe you can see that."

Elliott felt Howard sullied his reputation when Howard dressed him down in front of everyone on the day of the fire. It hurt his pride.

Duffus made his stance clear by replying, though in reality, his vote was necessary only to break a tie. Elliott hoped to draw support from Johnson, who'd brought a previous motion for Howard's dismissal. That one was narrowly defeated, but this time might be different.

"There's an election in a few weeks. It'll have to keep until the New Year."

From all other accounts, though, the efforts of Howard and crew were valiant. Given the resources at his disposal, no one else could have done a better job. Even if the firefighters understood fire behaviour and the chemistry of combustion better, under the circumstances, they were fighting a losing battle.

A.J. Mylrea stated in his report:

"The temperature in the interior of the building must have been very intense, for melted metal parts of machines may be found here and there in the ruins. In the front windows of the second story and in the front and some of the west windows of the third story of the concrete warehouse were found melted sash weights. In some cases they were but slightly fused, in other cases two or more weights had run together, and in several places on the third, shapeless masses of cast iron."

The terrific explosion rendered safety equipment in the factory useless in a heartbeat. Sprinklers fused, the water main into the plant severed, and fire separations blew off hinges. Pressure in the hydrants off-premise then plummeted. By the time other city staff got the pressure back up, firefighters had no chance to save the plant. Chief Howard had only sixteen firefighters to fight a blaze of that size. They would not have saved the plant, even with proper pressure at the outset.

Mayor Duffus visited New York City to consult with the Fire Chief on how they fought high rise and industrial fires. He came back, recommending the more powerful motor-driven pump that Chief Howard had requisitioned years ago. Council of the day voted to just repair what they had and hobble along status quo.

But by then, Council had relieved Chief Howard of duty. A Council who needed a scapegoat. The fire had cost them a commitment to spend over a million dollars to rebuild an improved Hunter Street Bridge, to keep the Quaker Oats in Peterborough. They fired Chief Howard in the summer of 1917. Council didn't even bother to wait until the British Fire Prevention Committee published their findings.

That the entire city didn't burn to the ground was a testament to the skilled acumen of the firefighters under the effective leadership of William Howard. With all that had happened, and lack of proper equipment, in no way, shape or form, was it reasonable to suggest Chief Howard was incompetent. He should have received a medal, but they showed him the door instead. Under his leadership, Peterborough Fire Department contained the fire to the Quaker property with a few exceptions, and saved the city.

When the published reports were available, they acknowledged the Fire Department did everything possible. The total loss was not a reflection of any error or negligence. Many experts agreed.

The Mylrea report commented on the rescue operations started after the Fire Department arrived. It remarked how quick actions of firefighters kept loss of life to a minimum. The authors of the report acknowledged the sensibility of beginning with rescue attempts. Quaker Oats can replace equipment and buildings, but not lives. Chief Howard and his brave team were heroes.

Before it was over, every available member of the Peterborough Fire Department worked for thirty-six continuous hours. Supported by a host of volunteers, including members of the former department and with help from both Lindsay, Lakefield and Toronto, they spared no exertion.

Speculation swirled around the future of the Quaker Oats Company's operations in Canada. The Quaker Executive Team was quick to confirm publicly it intended to rebuild. Before long, contingents from both the Town of Lindsay and City of Kingston, Ontario, approached the Stuart family. They came with attractive offers, hoping to lure the factory to their towns. Communities seeking opportunities, it would seem, are ever present in situations like this.

Kingston's effort to land the Quaker Oats Plant started only two days after the explosion, on December 13, 1916. Francis King, President of the Kingston Board of Trade, penned a communique to Robert Stuart and the Quaker Executive Team. In it, he touted the advantages of the City of Kingston and assuring the company of every support if Quaker Oats moved out of Peterborough.

After deliberating, Quaker Oats agreed to rebuild the plant on the existing grounds of the Peterborough location, though they had one condition. A larger, better bridge over the Otonabee River on Hunter Street. With business booming, more rail traffic was coming and going from the factory. The lower, smaller bridge made it difficult to increase capacity. Peterborough City Council agreed to this concession.

But a new factory took time to build, and the challenge of fulfilling existing contracts still posed a stumbling block. Leading into early 1917, Quaker Oats put a bold and creative plan into place. It harnessed unused capacity in local Mills and retooled other Quaker factories to satisfy the principal product lines produced at the Peterborough plant.

Production staff at the Peterborough location had the chance to move to another Quaker facility of their choice. Some took the company up on the offer, and moved to places such as Akron and Ravenna, Ohio. Perhaps they felt it was better to start fresh in a new place. Maybe they could not bring themselves to return to work at the Peterborough factory because of the terrible memories.

Overall, how the owners of the largest cereal company of the day handled this emergency was first-class from start to finish. Chicago executives dropped everything when they heard of the emergency and raced to what others might consider a back-water town. Throughout it, they showed leadership while accepting social responsibility to the City of Peterborough.

They did so without hesitation. Quaker Oats took an active role in the Fire Marshal's investigation and offered to pay staff wages for the month of December. The Stuarts announced a monthly allowance for the widows of the fallen and those permanently injured who could not return to work. Quaker Oats agreed to pay moving expenses and first months' rent at the new location for any who wanted to transfer to another city.

This was a company that truly cared for its staff and understood that company profits depended on those same hard-working employees.

Quaker Oats did not seek to maximize profits and cut corners to lay the burden for success on the backs of minimum-wage employees. They didn't cut and run when it best suited them, such as after a tragic loss like this one. It spoke volumes to the character of the Stuart family and the Board of Quaker Oats that they handled it as they did.

One hallmark of true leadership is to not shy away from a challenge; but to stand up and say: "*Something went wrong, and we want to understand that, then make things right again. And we're here with you, our sleeves rolled up, ready to do the hard work of fixing this.*"

The Quaker Oats Company exhibited true leadership.

Chapter Fourteen

In Memoriam

> *"But there was no need to be ashamed of tears, for tears bore witness that a man had the greatest of courage, the courage to suffer."*
> - Viktor E. Frankl, "Man's Search for Meaning"

One employee, interviewed by the local press after the fire, said that working at the Quaker Oats was *"... like a big family, that's what it was. The bosses were fine, and we were all so familiar with the rooms it seemed like a home. And the company was all the time putting in new machinery to speed up the work."*

The value of the property destroyed alone wasn't the only loss to the City.

This enormous family had just broken apart. It wasn't as simple as saying they only lost five percent of the company workforce. Each life touched many others. For those who remained to pick up the pieces, the cost was tremendous. Vivid nightmares haunted many of the survivors for years afterward. One stated not a day in years passed that she didn't think about the explosion.

It severed close friendships when employees picked up stakes and moved to the United States to work in another Quaker Oats factory. Add to that those who opted to find work elsewhere, because they suffered from shell shock. A term familiar to any who experienced war first-hand, but which also defined the aftermath of the Quaker fire. They didn't return to work at the plant because of fear. The entire city suffered the ripple effect of this tragedy.

Not every story has a fairy-tale ending. This one doesn't either. But there is a never-ending legacy of hope for a better tomorrow. And resilience. Life went on. Families picked up the pieces and moved forward. Quaker rebuilt the factory and Peterborough reimagined a beautiful bridge to span the Otonabee River on Hunter Street. Laura welcomed the newest member of the O'Brien family to the world on May 31, 1917. She named him Dennis, Jr., though family called him "Bud". Laura took in boarders at the home on Harvey St. to supplement the Quaker Oats payment. When the six children were a little older, she remarried, one of Dennis' first-cousins.

A story such as this cannot end, however, without paying tribute to those who fell as the result of the events of December 11, 1916. So many of their stories, already woven into the fabric of the pages of this narrative. Sometimes, little more than a name and a few brief details remain, the rest lost to the annals of time.

In the interest of respect, I honour their memories one more time. A person has not truly died, if we still speak their name:

Filippo Capone, Richard Chowen, John "Jack" Conway, Vincenzo Fornaro, James Foster, William Garvey, James William Gordon, Richard Healey, William Hogan, Walter Thomas Holden, Joseph Leo Houlihan, Edward "Ned" Howley, John Carter Kemp, Domenico Martino, Joseph Alphonse McGee, William Henry Mesley, Dennis M. O'Brien, William Miles, Patrick O'Connell, Thomas Parsons, Albert Ernest Staunton, William John Teatro, George Wellington Vosbourgh, William J. Walsh.

Every one of these men exhibited bravery and heroism on December 11, 1916. Unfortunately, little remains of their stories. In most cases, nothing more than genealogical statistics exist. Of those known, some are much more tragic than others.

But a life is more than birth or death dates, residence address, marital status, and the number of children one had. A life has depth and meaning. Each of the twenty-four names above represents someone's dad, son, brother, uncle, or friend. I only wish I could tell as much about each of them as I can of the few.

John Carter Kemp had a birthday approaching on December 28. He would have been sixty-seven. I doubt he wanted to celebrate that year, as his beloved wife Jemima had died on August 19. One of his daughters had passed away within the last five years. William Walsh sent John to the bottom of the elevator leg to check for fire. Doctors thought he might live, but he died of heart failure on December 14, 1916 at the Nicholls Hospital.

Kemp's bedside testimony was invaluable in helping Fire Marshal Heaton and the investigators put together a timeline of events. As revealing is what fellow Quaker employees said about John afterwards.

"*To him, his duty was as important as any work in the plant and he was diligent at all times in the discharge of his work.*"

John's job was to supply burlap bags to the men in the feed room who filled them with processed Vim Feed from the grinders above. One might say John didn't have a high-profile job. Yet he treated it as valuable as if he were the General Manager. Despite the physical work and personal tragedies, John Carter Kemp showed up every day with a smile on his face and optimistic about the future.

Walter Thomas Holden was one of many men to which John Kemp supplied feed bags to, for the duration of each workday. Holden's job, to sew up filled bags of livestock feed in the packaging department. When Kemp opened that access hatch and the area burst into flames, Holden must have been standing nearby because he too caught fire. Instinctively bolting into a full-on run, he sprinted towards an exit.

It would be hard to miss Walter, who witnesses described appearing as a human torch. A newspaper reporter commented he remained on the ground, stark naked except for the coats of several employees that covered him. Holden, scorched beyond recognition, was mistaken for

Edward Bedding, when a reporter asked someone nearby who it was. Walter Thomas Holden died of injuries in hospital on the day of the fire.

William J. Walsh was the Foreman in the Dry House. On the opposite end of that wall of flames from below, when he opened an access hatch on one of the attrition grinders, a detonation occurred. The energy of the blast blew Walsh out an upper floor window and deposited him at the Otonabee River. He blacked out and had no memory of anything after seeing flames.

Walsh's only injuries when found were a broken leg and jaw, and they expected him to make a full recovery. William developed a sepsis, though, followed by toxic shock, and died in hospital on December 27. Those who knew Walsh held him in the highest regard. He had the respect and affection of those who worked for him. Before he died, Walsh relayed what he witnessed to the investigating team, which helped in determining the cause of the fire.

Richard Chowen was the last man located alive in the building's debris. Badly burned, dehydrated, and suffering from hypothermia, they found him in shock. Chowen did not survive the ordeal and died on December 22. Before he passed, he provided an eyewitness account of those moments around the time of the explosion, which became invaluable to Fire Marshal Heaton's investigation.

Domenic Martino was a young man with an entire life ahead. He travelled to Canada from Italy for the chance of a better life. His wife and infant child remained back in Italy. Domenic sacrificed, scrounged, and saved enough to bring them to Canada. He had not been working at Quaker Oats very long, and only two-weeks before the fire sent money back so they could be together at Christmas.

Domenic was the only employee of the Boiler House to survive the north wall of the Dry House crashing through the ceiling, though mortally wounded. He succumbed to the injuries. Domenic's wife used the money sent for her own trip to Canada to repatriate his remains to Italy. Their first Christmas in Canada together, gone in the wisps of smoke.

Three souls have the heart-breaking distinction of having never been located or recovered from the factory: Jack Conway, Dennis O'Brien, and Albert Staunton. We may surmise that the intense heat of the fire vaporized their physical form and cremated them so completely that nothing remained to be found. Conway came down to tell Walsh he smelled smoke and the first explosion likely took him.

Witnesses have O'Brien disappearing when the floor below him gave way. If two floor assemblies had sandwiched him, rescuers would have found Dennis, the concrete shielding his body from the heat. It is more plausible that when the floor disappeared, Dennis plunged into a blast furnace of such severe temperature that it obliterated him with deadly precision.

Prior to that, several co-workers saw Dennis re-entering the burning building multiple times. The number of times varies between three and five. No matter how many occasions, O'Brien became one of heroes of that day. More than a few people owed their lives to Dennis' bravery, but a small recompense to Laura and the children. They not only lost a father and husband, but a grandfather in one fell swoop.

Vernon's City Directory of 1916 lists Albert Staunton as a shipper. One may presume Staunton worked in the Shipping Department, in the Warehouse Area, which did not see fire for at least a half hour. Staunton, under normal circumstances, had ample opportunity to escape to safety. That he did not survive leads a reasonable person to conclude that, like Dennis O'Brien, Albert Staunton sacrificed his own life to save others. He, too, was a hero.

For families of the three, horrible enough that search crews didn't recover their bodies; but also not knowing what happened. Worse, the Church didn't allow a Mass of Christian Burial, since there had to be a corpse. If any had life insurance, it took a few years to make a claim to receive the proceeds in the absence of a body.

It is staggering to think of William Mesley and family. All six daughters worked at the factory that morning. Thankfully, each escaped serious injury, but they knew what part of the plant where their father worked.

The destruction of the Boiler House left little optimism that their dad survived. When he did not turn up at either hospital that afternoon, it broke their hearts. It is an irony, or a twist of fate perhaps, that just months before he'd been working at Bonner-Worth Mill. His family, in hindsight, wondered if he should have left the Mill. They wouldn't be all together as a family for Christmas, as he'd hoped.

Likewise, consider William Garvey. He survived the horrific injuries sustained that morning, and lived for another two years, passing away in November 1918. He moved to the packaging department for a promotion to assistant foreman. Weeks before, he worked in the Dry House, along with many who survived the explosion. Garvey's story is a prolific example of resilience.

Thanks in part to the brave efforts of brother Frederick and Edward Skilleter, William Garvey recovered enough that he returned to work full time. Quaker Oats promoted him to Foreman of the Cereal Department. The Garvey's added another child to their family. On the surface, this may be a good news story. But, in the Fall of 1918, Garvey contracted influenza, which became pneumonia around October 24, and within ten days, he had succumbed to the illness.

It's easy to suggest Mr. Garvey became another victim of the Spanish Flu. He might have fought it, if not for a compromised immune system, thanks to the severe burns sustained on December 11, 1916. Goodness knows that inhaling smoke and fire for so long before being rescued damaged his lungs. That he lived in the first place was miraculous. William Garvey left behind his wife, Mary, and five small children at home.

One of the last tragedies, Richard Healey. Somehow, he endured the massive skull fracture received that morning, and "dying" three separate times in hospital in the coming days. Healey lived another twelve years and passed away in 1928. Richard never returned to work, and with seven young children at home, wife Adela struggled. The oldest, Leo, aged eleven, had to drop out of school to help support the household.

Richard's remaining twelve years were hard. He suffered brutal headaches daily and needed constant care. In an era when the man was head of household, he surely felt helpless, if not useless. Yet, as with others, Healey exhibited resilience and courage in the face of adversity. The family carried on. They had no other choice. Richard Healey was just forty-four years old when he died.

But that is less than half of the men who died, and a fraction of those injured. When compounded to add those whose lives the tragedy permanently altered, the true cost of this event is substantial.

Epilogue

My maternal grandfather, George O'Brien, was born on May 7, 1909. I gave so much thought in researching and writing this narrative about the sheer adversity he witnessed in one lifetime, a single generation. At five years old, the Great War broke out. By the time it ended, it wiped 22 million people off the earth. At age seven, the events described in this book transpired, and George lost his father, Dennis and grandpa, William. The First World War ended when he was nine years old and then came the Spanish Flu.

The pandemic that followed infected an estimated one-third of the world's population and killed 50 million people. George was only ten years old. The year he turned twenty, the Stock Market crashed on October 4, 1929, sparking a global economic crisis. Like many others, he spent the entire decade of the 1930s enduring the Great Depression. High unemployment, inflation, and famine characterized the era. An event of such magnitude, no one in the world remained untouched by it.

At age 24, as the world's economy crept out of depression, the Nazis come into power. World War Two broke out when my grandfather was thirty years of age. Everyone heard about the dropping of an Atomic Bomb on Hiroshima and Nagasaki in 1945. Prior to that, the bombing of Pearl Harbour and stories of the internment of Japanese Americans. George just turned thirty-six. News began emerging of the horrific atrocities of the Holocaust while the Allies fought to stop Hitler. Another 60 million died in that six-year period.

After a brief period of peace and economic prosperity following World War Two, the Korean War broke out in the early 1950s, followed a decade later by the Vietnam War. Closer to home, people of colour fought for Civil Rights, and social unrest existed because of both segregation and in protest of the war in Vietnam. Violent uprisings marked this period as Black Americans fought for the equal rights, just as the US Constitution had promised.

The 1960s represented a turbulent time in North America. Even if the average citizen of Peterborough, Ontario stayed isolated from much of that, it was in the general consciousness of all. John Kennedy was assassinated on November 22, 1963; Malcolm X on February 21, 1965; Martin Luther King, Jr on April 4, 1968; and Bobby Kennedy on June 6, 1968. George lived through these historic events of the 1960s, including ones less controversial such as Neil Armstrong's famous lunar walk on July 20, 1969. George O'Brien hadn't even attained retirement age by the close of that decade.

I learned so much from my grandfather in the brief time I knew him. But he never spoke to me of what happened at the Quaker Oats on December 11, 1916. I didn't understand the many adversities faced in the aftermath and his lifetime. I simply remember a gentle man who had a love for Coca Cola and ate bacon and eggs most days for breakfast.

A devoutly religious man, George O'Brien was wise and preferred to do things in his own way. He challenged my own way of thinking in positive ways during those formative teenage years. I suppose we bear our own crosses. His, the loss of Dennis. I see now he bore it with honour and never recall him being defeated. And now that the book is done, I fancy him as my connection with Dennis, whom I knew very little about before 2016.

While I never physically met my great-grandfather, Dennis O'Brien, I sense I did meet him intangibly. A piece of him existed within George. If he lived past the fortieth year, perhaps Dennis would have aged to be gentle, too. A man who challenged his grandchildren positively to contemplate the world and their place within it.

I doubt a single person alive today has seen so much change and strife in their lifetime as anyone born in that first decade of the 1900s. And frankly, I'm not sure if I possess the intestinal fortitude to have survived. I live a sheltered life and, notwithstanding current affairs in the world at the time of this writing, I consider myself fortunate to not have had to endure such instability. Because of my heritage, I am privileged.

A child born in, say, 1985 believes their grandparents just don't understand how tough life is, that they cannot relate to how hard things are for today's youth. Today, we possess the comforts of a modern world, with a brand-new pandemic. Yet complain because we must stay confined in our homes where ample food, running water, electricity, Wi-Fi and even multi-media sources of entertainment exist to while away the hours in lockdown. None of these things existed back in the day.

Humanity of a hundred years ago endured much and yet survived tragedy without losing a zest for living. Life was trying, yet they stood in the face of it and persisted. A slight change in perspective can generate miracles. We could take a lesson from our ancestors in doing everything possible to help and protect one another and live out a sense of community.

So, what does any of this have to do with the story I just wrote? Everything. First, because I cannot stress enough how the events of December 11, 1916 in Peterborough, Ontario were not just a "one-off" of misfortunes witnessed by that generation. Second, to acknowledge this as yet another shining example of resilience and courage in the face of adversity that those ancestors and many others showed, to which I am proud to share with you.

Matthew Flagler
Peterborough, Ontario, Canada
November 2020

Quaker Oats

Partial Staff List 1916

In December 1916, there were approximately 500-550 men and women employed at the Quaker Oats Company in Peterborough. It was about a 50/50 split between men and women. Finding a complete staff list was trying, to say the least. I first contacted what is now Pepsi Quaker Tropicana Gatorade, or Pepsi QTG for short. Their response was that staffing lists were confidential and could not provide me with this information.

My next obvious source was Vernon's City Directory for Peterborough 1916. Back in the day, the Vernon Company, in Hamilton, Ontario produced directories of citizens for every major (and many minor) cities and towns across Canada. They produced these annually and made the directory available for sale in stores and shops in each city.

They published their directory in a time long before Ma Bell provided a telephone book for free. Most people in 1916 in a place like Peterborough did not have a telephone in their home. The Directory had three parts. The first was a listing of citizens' residences and business locations by street name. So, each heading would start with the street name, and then list all the houses by number on that street and the head of household living in that house.

The second section was an alphabetical listing of residents, and beside their name it would say their address and often their employment. It didn't always list the name of the company, sometimes it was just a job category such as "driver", "teamster", or "soldier". If a woman was widowed, it would say "widow" and then name her former husband. The last section was a municipal and business directory, where one might find the address and phone number of City Hall, the Fire Department, or the local butcher.

The enormous challenge in compiling this staff list was in levels. First, they produced the guide months before the new year to be available for purchase in January of that year. The period to which we are discussing saw people occasionally changing jobs or moving in or out of an area. So, for example, the Vernon City Directory for Peterborough 1916 as published shows Percy Naish, our lucky man who survived three major catastrophes unscathed, as working as a salesman at JC Turnbull Company.

He likely ended 1915 or began 1916 as a salesman at the Turnbull Building, but we have multiple other sources placing him in the factory at Quaker Oats on December 11, 1916, and also where in the plant he worked. We can rely on these historical sources more so than the Directory. If it wasn't for those other sources, one would never know that Naish worked at Quaker Oats that December morning. And such as it is with many others, no doubt.

Next, the Vernon Directory only lists those who lived in the City of Peterborough, and excluded the outlaying areas of Smith, Douro, North Monaghan, etc. So, one would not find the name of William Hogan in the directory, as he was only staying temporarily with Dennis and Laura in December. There were surely other employees who lived outside the City limits and travelled back and forth into work each day. I could not find them.

Then, regarding the City Directory, we have the fallible human ability to make mistakes. In doing my research, I found spelling errors and things out of alphabetical order. It was not common, and the Directory was an excellent resource. Also, it did not always mention the name of a person's employer. So, our janitor, Edward Skilleter, they listed as being employed as a janitor but not at Quaker Oats.

Nonetheless, I opted to take the time-consuming approach of scouring through the directory for information and have come up with the list that follows. It numbers around 220 employees. Most of those are men. Again, unfortunately, the guide didn't list many young single women, especially if they were boarding somewhere like the YWCA or living with their parents. The directory listed the name of the husband but not always spouse or children. This was the unfortunate drawback to a male-centred society of 1916.

My last source of information was the research into the tragedy. Many names of employees showed up in the fire investigation, newspaper reports, and other publications about the fire and explosion. For example, the Vernon Directory had none of the Italians who died in the fire listed in it. But we know they worked there because the local hospitals listed them as patients; and later among the list of those who died.

It was tedious enough work to compile the list that I have, and I doubt I could ever provide you with a complete list without the help of the company itself. I apologize in advance that the list is predominantly men. That had nothing to do with favouritism, but for the reasons mentioned above.

Allen, James
Allen, Michael
Allen, Robert, engineer
Armstrong, James
Armstrong, Wilbert
Aspell, Ralph
Axford, Neil
Bedding, Edward
Bettes, John H
Betts, Sidney
Bissonnette, Fred
Blodgett, Ambrose
Booth, Eva
Booth, Irene F. clerk
Briggs, Robert, clerk
Brooks, William J., clerk
Brown, William F., clerk
Burton, George
Caccavella, Michael
Caccavella, Fred
Campbell, Thomas
Capone, Fillipo
Carr, Edward T., sales correspondent
Carruthers, George C., foreman
Chandler, Harry
Chowen, Richard
Chrow, William
Clancey, John
Clancy, James J., Jr, clerk
Clegg, Charles E., foreman
Cocks, Thomas
Collins, Joseph
Conlin, Augustus M.
Conway, John "Jack"
Corkery, Stephen
Cunningham, John J.
Curran, John
Curran, Patrick
Curtin, Joseph
Dalton, Percy
Dalton, Thomas
Davey, Ernest
Davidson, George
Davis, George
Davis, Norman S., clerk
Denham, William, Superintendent
Denne, THG, Sales Mgr.
Dinnen, Timothy
Dixon, Alex
Donoghue, Eileen, switchboard
Drain, William
Dwyer, Wilfred
Edwards, George P., Ass't Superint.
Everson, Arthur, foreman
Febbo, Nicholas
Feeley, Leon
Fenn, David
Ferguson, George
Fife, Herbert A., Chief Electrician
Fornaro, Vincenzo
Foster, James
Francis, John H.
Franks, Ferdon W., clerk
Freeman, John A.
Frost, George
Frost, William C., Foreman, oatmeal

Gandy, Fred J., foreman
Garvey, Frederick James
Garvey, William John
Giardini, Mauro
Gill, Alfred
Gilman, Robert
Gordon, James W.
Gorham, Micheal
Grant, Thomas J., foreman
Griffin, James W.
Guerin, Dennis
Guerin, Frank
Guerin, Martin
Hamilton, Bert
Hanrahan, Thomas
Hardwick, George T.
Harran, Walter
Harrison, Thomas
Hartley, Percy
Healey, Patrick
Healey, Richard
Hennessey, Herbert
Herd, William
Hickey, Thomas
Hines, George
Hockey, Fredrick C.
Hogan, William
Holden, Walter Thomas
Holland, Pearl
Hoolihan, Dennis
Houlihan, Joseph Leo
Howley, Edward "Ned"
Hunter, Roland W, clerk
Jamieson, Amy B, stenographer
Jardine, Carmen
Johnston, Harry, electrician
Jones, Edgar
Joyes, Charles
Kearns, John
Keenan, William, checker
Keir, David
Kemp, John Carter
Kennedy, Daniel
Kenyon, Olga, stenographer
Killoran, Matthew
Kirn, Howard, clerk
Kirn, Martin, chemical engineer
Knight, John, fireman
Kylie, Emmet
Lashbrook, Albert
Logan, Earl E., clerk
Long, Michael
Lush, Gilbert
McCarrell, James
McCarroll, Mellville P., clerk
McCarthy, Joseph
McCloan, Barney
McCoy, Lewis
McGee, Joseph "Alphonse"
McGrath, David
McIlwain, William T.
McKee, Clarence
McKeown, Harry F., millwright
McLaughlin, Robert
McLeish, Arthur JD., clerk
McMahon, Thomas

Malane, Fred J., clerk
Marock, Robert
Marshall, Alfred
Marshall, George
Marshall, James
Martino, Domenico
Martinson, George
Masters, John
Mead, James, fireman
Mesley, Edna
Mesley, Ellen
Mesley, Fred, engineer
Mesley, Gladys
Mesley, Leitha
Mesley, Lula
Mesley, Marjorie
Mesley, William Henry
Metherall, Wilmot
Milaney, Luke
Miles, William
Miller, Alex, clerk
Miller, James, watchman
Minicolo, P.
Mongraw, James
Moore, Ira D., sls corr
Morphet, Vernon E., clerk
Murphy, James
Murphy, William
Naish, Percy
Nicholls, Daniel, laborer
Nolan, Laughlin
Nolan, Thomas E.
Northey, James
O'Brien, Dennis M.
O'Connell, Patrick J.
O'Connor, Frank
O'Connor, Patrick
O'Connor, William
O'Donnell, William
O'Toole, Stephen
Packenham, James, millwright
Parks, Charles
Parsons, Thomas
Payne, Albert
Payne, Edward
Payne, Hilliard E., clerk
Pearson, Gordon
Perdue, William E., foreman
Pidgeon, Alex, clerk
Pomeroy, William, electrician
Powell, Conrad
Power, Mabel
Primeau, Alex, bookkeeper
Pulver, Samuel
Quinlan, Thomas
Ray, George
Rea, Thomas
Robins, L.
Rowe, James H.
Sanderson, Fred, shipper
Sant, Thomas
Schnick, John A., miller
Scrivens, Frederick
Searight, Edwin, clerk
Seely, Herbert, miller
Serro, Michael, foreman

Sheehan, Emmett, bookkeeper
Simmons, Alfred C.
Simmons, Nelson O.
Simpson, Andrew
Skilleter, Edward C., janitor
Smith, George
Staunton, Albert
Stant, Thomas
Staples, Newton, clerk
Steadman, James E.
Stewart, Henry, miller
Storey, Fred
Stuart, Robert, inspector
Sturmer, Charles
Sullivan, John E., cashier
Tangney, Eva M., stenographer
Teatro, William John
Thackeray, A.E. Clark, clerk
Thackeray, William N.
Tobin, Michael
Todd, John
Tougas, Joseph
Trethewey, Albert, millwright
Vosbourg, George Wellington
Walsh, William J, foreman
Weir, John R., shipper
Welch, Charles
Williams, Alfred
Willshaw, George T., packer
Withers, Cyril
Young, Fred
Young, Matthew

Peterborough Fire Department circa 1916
Southwest corner of Aylmer & Hunter Streets

St. Joseph's Hospital Peterborough
Circa 1916

Photo taken day of the fire, December 11, 1916
East side of factory.

Another photo taken in the aftermath of the fire.

Photo from inside the Quaker Oats Factory early 1900's.

Interior Factory Photo

Interior Quaker Oats Factory Peterborough

Quaker Oats plant, c. 1915.

Dennis O'Brien

Vincenzo Fornaro

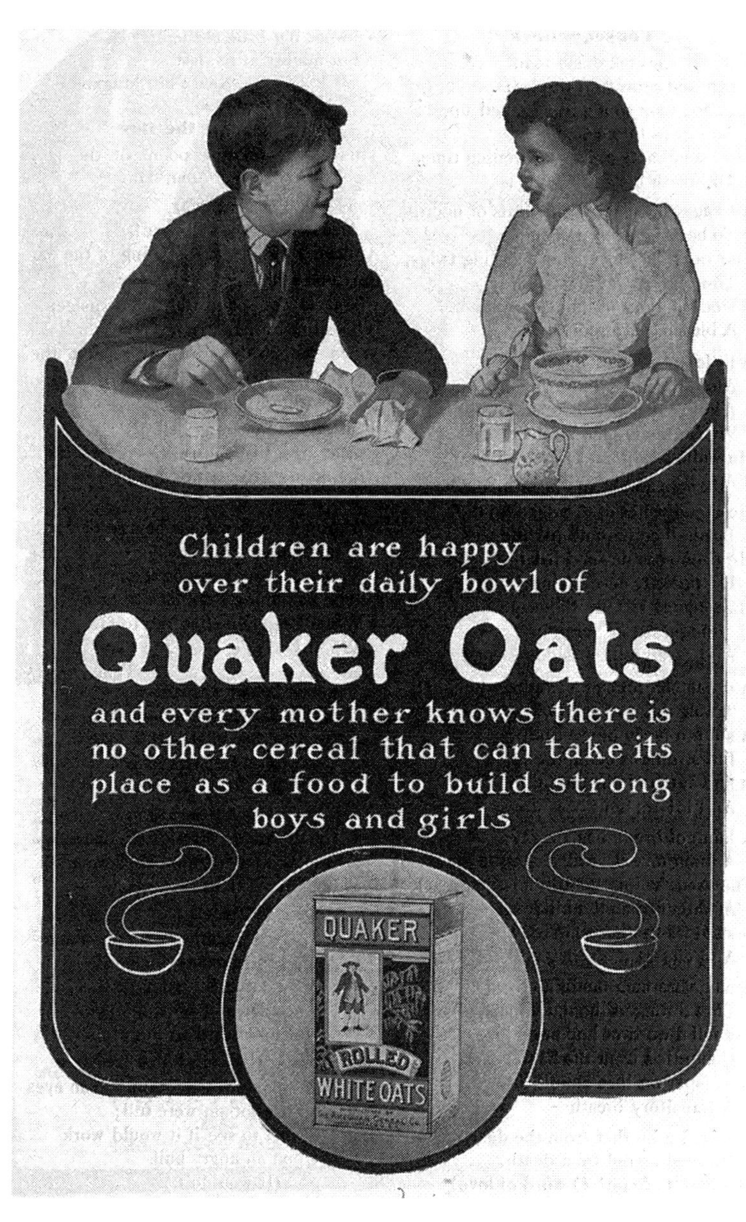

Quaker Oats Advertisement circa 1905

Quaker Oats Advertisement

MEMORIAL PLAQUE PLACED IN MILLENIUM PARK, PETERBOROUGH, ONTARIO ON THE 100TH ANNIVERSARY OF THE TRAGEDY.

Manufactured by Amazon.ca
Bolton, ON